Science in Today's Classroom

Science in Today's Classroom

First Edition

EDITED BY C. Sheldon Woods

Northern Illinois University

Bassim Hamadeh, CEO and Publisher
Angela Schultz, Senior Field Acquisitions Editor
Michelle Piehl, Senior Project Editor
Casey Hands, Associate Production Editor
Jess Estrella, Senior Graphic Designer
Stephanie Kohl, Licensing Coordinator
Don Kesner, Interior Designer
Natalie Piccotti, Director of Marketing
Kassie Graves, Vice President of Editorial
Jamie Giganti, Director of Academic Publishing

Copyright © 2020 by Cognella, Inc. All rights reserved. No part of this publication may be reprinted, reproduced, transmitted, or utilized in any form or by any electronic, mechanical, or other means, now known or hereafter invented, including photocopying, microfilming, and recording, or in any information retrieval system without the written permission of Cognella, Inc. For inquiries regarding permissions, translations, foreign rights, audio rights, and any other forms of reproduction, please contact the Cognella Licensing Department at rights@cognella.com.

Trademark Notice: Product or corporate names may be trademarks or registered trademarks and are used only for identification and explanation without intent to infringe.

Cover image: Copyright © 2017 iStockphoto LP/kali9.

Printed in the United States of America.

CONTENTS

1 Foundations of Science Education 1

    1. Constructivism and Science Education 2
  By Claire T. Berube

    2. A Case for Culturally Relevant Teaching in Science Education and Lessons Learned for Teacher Education 37
  By Felicia Moore Mensah

QUESTIONS TO PONDER 57

2 Basic Tools of Science Education 59

    3. Basic Science Process Skills 60
  By John Settlage and Sherry Southerland

    4. The 5E Learning Cycle as a Model for Science Teaching 94
  By John Settlage and Sherry Southerland

    5. Let's Experiment! 123
  By Ann Finkelstein

    6. Science, Technology, and Society 130
  By Ronald V. Morris

QUESTIONS TO PONDER 151

3 Science Education and Societal Factors 153

    7. Differentiating Science Pedagogy 154
  By Anila Asghar

    8. Science Curricular Materials through the Lens of Social Justice 173
  By Mary M. Atwater and Regina L. Suriel

    9. The Brains-On Approach to Science 184
  By Ann Finkelstein

QUESTIONS TO PONDER 192

4 Science Education and Science Learners 193

10. Fostering Scientific Reasoning as a Strategy to Support Science Learning for English Language Learners 194
    By Cory Buxton and Okhee Lee

11. A Neglected Dimension of Social Justice 218
    By Mary John O'Hair and Ulrich C. Reitzug

12. Assessing Learning From Inquiry Science Instruction 233
    By Stephanie B. Corliss and Marcia C. Linn

13. More Basic Science Process Skills 257
    By John Settlage and Sherry Southerland

QUESTIONS TO PONDER 284

Section I

FOUNDATIONS OF SCIENCE EDUCATION

Objectives

- Analyze some of the factors that affect contemporary science education.
- Determine the role constructivism plays in a learner to develop scientific literacy.
- Explain why constructivism and multiculturalism are important in today's science classroom.
- Develop a rationale for multicultural science education.

Key Concepts

constructivism: An epistemology and learning philosophy popular in contemporary science education where the learner is an active participant in building knowledge.

multiculturalism: An approach that seeks to eliminate prejudice and bias of any type, conscious or unconscious, individual or institutional, that serves as a barrier to the survival and self-determination of individuals and communities.

Introduction

This section of the book is meant to give you a foundation for the teaching of science in an elementary classroom. Readings that cover the basic topics of constructivism and culturally relevant science teaching have been compiled in this volume. While these topics are not the only ideas that are important, they are critical considerations for changing demographics in America.

READING 1

CONSTRUCTIVISM AND SCIENCE EDUCATION

By Claire T. Berube

John Dewey never used the word "constructivism," nor did anyone else of his day. For at least a century before the term "constructivism" was coined, educators were implementing the philosophies and practices of the constructivist education movement under other names and philosophies. The leaders of the progressive movement spawned many such outgrowths, including social justice theory, constructivism, child-centered learning, and discovery learning. Today, constructivist learning philosophies drive not only many school districts' curricula and teacher practices, but many college and university teacher preparation programs as well, including some of the leading universities in the nation. It is fraught with controversy and disagreement among educators the world over, but it serves as a valid, highly effective model for educating the nation's children.

The idea that children build knowledge from their own experiences and mode of thought is the concept behind constructivism. It is the most important component of science discovery learning also. Another way of thinking about it would be that children and people in general, tend to extract different things from the same lesson or experience, depending on their world views and experiences prior to the learning. The coining of the term "constructivism" can be traced back to Piaget's reference to his views as "constructivist" and from Bruner's description of his discovery learning technique as "constructionist" (Applefield, Huber, & Moallen, 2000–2001). Those employing constructivist methodologies believe that real understanding occurs only when children participate fully in the development of their own knowledge, which occurs morally, cognitively, mentally, and socially.

> They describe the learning process as self-regulated transformation of old knowledge to new knowledge, a process that requires both action and reflection on the part of the learner ... the research of cognitive psychologists and science educators over the past decade has shown that what children learn greatly depends on what they already know. Knowledge and understanding grow slowly, with each new bit of information having to be fitted into what was already there. (Howe & Jones, 1993, pp. 8, 9)

Constructivism is also a philosophical explanation about the very nature of knowledge itself. As an epistemology, constructivism declares that knowledge is formed by the knower from existing beliefs and experiences. Knowledge is not independent of the knower and is not made up of accumulated "truths." Individuals create their own meaning from their own experiences; therefore, all knowledge must be tentative, personal, and subjective. Also, constructivism is an epistemological view of knowledge formation emphasizing construction rather than transmission and recording of information given by others (Applefield et al., 2000–2001).

Constructivism also can be defined as programs that are student-centered and are based on a theory of learning that focuses on how students develop understandings (Richardson, 1999). The constructivist approach differs from the traditional (direct instruction) approach in that students are included in the learning. Teachers who instruct from constructivist pedagogy develop lessons that lead children to engage in self-directed problem solving instead of direct instruction. Science classrooms which are taught by teachers who only lecture and "tell" are not conducive to student learning at the highest level. Science, more than any other subject, demands teaching styles that force the students to think for themselves, question conformity, and create their own learning.

> Most constructivists would agree that the transmission approach to teaching, usually delivered through lecture or direct instruction, promotes neither the interaction between prior and new knowledge nor the conversations that are necessary for intense involvement in ideas, connections between and among ideas, and the development of deep and broad understanding. (Richardson, 1999, p. 146)

Teachers assess the prior misconceptions that students bring to the classroom and try to correct them through this identification. Students use hands-on and cooperative learning situations and lessons that are student-centered based on children's basic curiosity about the world.

Also, constructivism is concerned with linking students' prior knowledge to present activities. According to McNichols (2000), "Constructivism is a theory about knowledge and learning." Embedded in this theory are the notions that:

1. Meaning, which is represented as knowledge, is based internally in the learner.
2. The acquisition of knowledge is the responsibility of the learner.
3. Knowledge is achieved from the learner's experiences and values conditioned by reflection, inquiry, and cognitive dissension.
4. Learning is an internal process, which is enhanced through the consensual negotiation of ideas.
5. The outcome of knowledge is a pragmatic process.
6. The assessment of learning is naturally connected with the learning process (McNichols, 2000).

These tenets of constructivism imply a classroom setting where social and intellectual interaction help students form meaning of the subject matter. In the science classroom, cooperative groups are the best way that students can tackle problems together to devise several solutions to one problem. Thus, constructivist pedagogy does not direct teachers in what and how to teach, but urges instructors to facilitate learning by providing a conducive environment for such in the classroom. Of course, many things go into successful teaching. There are old fashioned teachers who use lecture as the majority of their instruction, and who are very good at delivery and story telling, so that the students' attention is held. But again, in the science classroom, this is the least effective teaching technique. What constructivism seeks to add to the classroom experience, is the child in the role of his or her own educator, with the teacher as the guide. This does not take the teacher off the hook as the leader in the classroom. Nor does it ask of the children to learn the material themselves. Rather, instead of telling the students the answers that the teacher already knows, the teacher provides problems for the children to solve, thereby stressing process over product, and solidifying the learning in the children's minds. Things are better remembered when they are discovered and "worked out," rather than being passively received.

One of the biggest criticisms of constructivist pedagogy is that teachers are thought to have less content knowledge than teachers who use a more direct instruction approach. They are thought to "let the children teach themselves," which could not be farther from the truth. The best constructivist teachers also possess the highest levels of content knowledge for their discipline. This however, *can* be a serious problem in an elementary and middle school science classroom. The average elementary and middle school science teacher is a woman, and most women are uncomfortable with science as a topic to teach.[1] I will discuss this in depth a little later, but it is worthy of discussion.

COMPONENTS OF CONSTRUCTIVISM

In order to understand constructivist practices in terms of their origins in psychology and educational philosophy, it is necessary to separate them into components, along with their corresponding research studies. The components that this book will address are concept formation, cooperative learning, alternative assessment, hands-on/active learning, and student-centered learning.

Concept Formation

Lev Vygotsky (1896–1934) was a Russian psychologist interested in the formation of language and thinking and cognitive development. Vygotsky stated that one of the basic components of constructivist pedagogy is the notion that children develop concepts on

---

1 This perspective does not reflect the viewpoint of the editor/publisher.

their own through everyday experience, called everyday concepts, and those concepts learned in school, called scientific concepts. These scientific concepts may be remote from a child's experience unless a teacher knows how to tie them into the child's experiences to make them meaningful. Conceptual change is the term that refers to the ongoing process in which children integrate their everyday concepts into a system of related concepts, including scientific concepts that have been taught in school (Howe & Jones, 1993).

Vygotsky, more than any other philosopher other than Dewey, had a huge impact on constructivist science classrooms. Vygotsky contended that students learn better through social interaction, since discussions and feedback take learning to higher and higher levels. Nowhere in education is this better represented than in a science lab with cooperative groups. Problem solving, hypothesizing, and scientific discovery demand different viewpoints and the cooperative group provides this to students. Also important is that in the real science world, experiments do not always lead to great discoveries. Science students must learn that there is value in narrowing down possible solutions until the correct one is found, and this concept of conceptual change is necessary to move from one idea of possible solution to another more valid solution. (Conceptual change theory will be discussed in more detail later in this chapter.)

The following instructional techniques help to accomplish this goal.

Reciprocal Learning

Ongoing dialogue between student and teacher is at the heart of constructivism and helps to prevent student misconceptions of learning. To gain new understandings from one's social environment and to become a high level thinker capable of making meaningful connections requires adopting specific intellectual skills that are modeled by competent teachers. Learning-to-learn strategies may be taught to students or discovered by students as they attempt to solve problems. Reciprocal teaching is one such strategy (Applefield et al., 2000–2001). Reciprocal learning and teaching strategy is the creation of Palinscar, David, and Brown (1984), but based on the work of Vygotsky. It is a strategy employed in order to raise reading comprehension, which includes four points:

1. Summarizing
2. Questioning
3. Clarifying
4. Predicting

The procedure consists of interactive dialogue where the teacher models the four skills, gradually letting the students take over the responsibility, while taking the role of coach. The teacher and students take turns leading a dialogue concerning sections of a text. They also take turns generating summaries and predictions and in clarifying misunderstandings in the text. The order in which the four strategies occur is not important, most teachers mold the four to the particular text being read (Palinscar et al., 1984). The goal is to encourage student regulated self-learning by helping students develop effective strategies and contextual knowledge of when to use them (Applefield et al., 2000–2001).

In research studies conducted by Palinscar et al. (1984), students increased their comprehension ability after receiving reciprocal teaching instruction, including modeling and corrective feedback on the four comprehension activities. The types of tasks selected for students included complex, real-life problem-based tasks, which emphasized conceptual understanding over memorization (Applefield et al., 2000–2001). Empirical support for reciprocal teaching technique is found in several comprehension studies (Palinscar et al., 1984, 1992), and results confirmed that the reciprocal technique can build pre-reading and comprehension skills (Andrews, 1985).

According to Palinscar, David, and Brown (1984), the goal of long-term reading instruction is not to focus on content knowledge that students to a large part already possess, but to stress comprehension-fostering strategies that extend knowledge to more areas other than reading. In a study conducted in 1984, teachers received training in reciprocal techniques for a reading class and students were measured on criterion tests comprehension, reliable maintenance over time, generalization to classroom comprehension tests, transfer to novel tasks, and standardized tests. These measures also were taken from traditional classrooms with no intervention. Reciprocal teaching techniques accounted for significant gains in each of these measures. Many of these results were replicated during a second study.

Reciprocal learning improved listening comprehension as well. In a study conducted at the primary level to determine whether reciprocal teaching would be an effective approach to improve nonreaders listening comprehension, before the administration of the treatment (reciprocal teaching), pretest scores were 51% correct for the reciprocal group against 49% correct for the traditionally taught group. After treatment, posttest scores were 72% for the reciprocal group against 55% for the traditional group. Reciprocal teaching was compared to traditional basal reading instruction where both sets of students read the same text from basal readers (Palinscar, David, & Brown, 1992).

Reciprocal learning theory has as its foundation Vygotsky's learning theory. Vygotsky's unique ideas about education and socialization of children which are relevant to science teaching were developed through observing children going about their daily business of school, family, and play, and emphasized the importance of interactions with others as it fosters cognitive development. Vytogsky emphasized the role of guided learning in social contexts, which is the basis of reciprocal learning (Palinscar et al., 1992) and the bedrock component of today's constructivism. Vygotsky's contribution to constructivism has been identified with social constructivism because it emphasized the importance of social context for cognitive development.

Vygotsky's best known concept in the social context is called the zone of proximal development, which could be another term for reciprocal learning. It argues that students can, with the help of teachers and slightly more advanced students, master ideas and concepts that they could not master by themselves. He believed that

> children should have tasks set for them that are just beyond their present capability but which they can perform with guidance from a teacher or more advanced peer. He described a "zone of proximal development" (ZPD), as an area just beyond a child's current level of ability. (Howe & Jones, 1993, p. 31)

As a middle school science teacher, I tried to teach in this "zone" every day. I would pose difficult problems, but provide the metaphoric ladder that the students could use to find the answer. And if I saw their frustration growing after a time, I would throw them a portion of the answer until they got it. I provided a scaffold for their learning.

As mentioned, Vygotsky's concepts are aligned closely with science education. Successful science classrooms stress cooperative learning (as mine did), with laboratory experiments serving to enhance social skills and cooperation in the completion of science process and lab skills. In addition to Vygotsky, this style of teaching has at its foundation the theories of Dewey, Piaget, and Bruner (Howe & Jones, 1993). There are four general principles that are applied in any Vygotskian classroom:

1. Learning and development is a social, collaborative activity.
2. The zone of proximal development can serve as a guide for curricular and lesson planning.
3. School learning should occur in a meaningful context and not be separated from learning and knowledge children develop in the "real world."
4. Out-of-school experiences should be related to the child's school experience (Howe & Jones, 1993).

Vygotsky has filled in gaps some scholars find in Piaget's work, such as not including the importance of social dimensions and their influence on intellectual development. Vygotsky's theory suggests the inherent social nature of all humans and his work marries social with intellectual instead of divorcing the two. Socially mediated instruction as it pertains to Vygotsky is called scaffolding. The nature of scaffolding is for the teacher to provide enough support without doing the work for the student (Palinscar et al., 1992).

Albert Bandura (1977) has also studied human behavior in a social learning theory that he calls "reciprocal determinism." In this theory, human behavior influences environment and environment influences human behavior. People and environments do not function independently of each other, rather they determine each other. This is the opposite view of behaviorism which states that a stimulus always causes a response; a one-way directional relationship. Behaviorism neglects determinants of behavior caused by cognitive functioning. Social learning theory relies heavily on self-regulating capacities within the individual, thereby placing some responsibility on the person and not solely on the stimulus. In the constructivist classroom, this would have implications for students who interact and participate in their learning rather than experiences a more passive learning experience.

The Learning Cycle

Constructivism is based on the notion that students build knowledge by continually restructuring new information to fit existing concepts. The learning cycle is a conceptual-change model of instruction that is consistent with concept formation. It has several components that are similar to reciprocal learning. The three-stage model is as follows:

1. Exploration phase—teacher gives students materials and encourages exploration and questions about things dealing with new materials that they do not understand.
2. Concept introduction phase—teacher introduces and explains key concepts, may illustrate, diagram. Textual readings become more purposeful.
3. Concept application phase—teacher help the students apply the newly learned concept to new situations.

The learning cycle is based on the work of Piaget and his learning principles of mastery and self-regulation, where learners develop new reasoning patterns as they accommodate and assimilate new ideas. Students become reflective and as they practice new skills, they improve their cognition rather than their behavior as in the case of behaviorism, which is what drives the traditional teaching method (Ebenezer & Haggerty, 1999). Employing the learning cycle also clarifies students' thought processes and misconceptions. Students have the opportunities to explain and debate their ideas, thereby giving teachers good insight as to why students are arriving at certain answers or viewpoints (Bevevino, Dengel, & Adams, 1999). The learning cycle is crucial to science education because it brings the student back time and time again to the basic question: does the data support a new idea that suggests that I have to change my old beliefs? Much of the content in science is intimidating to students, and it takes teachers willing to train students to question belief systems in order to move on to the next, higher level of understanding. Much of learning can be blocked by students old belief systems holding them back.

Musheno and Lawson (1999) studied to see whether the learning cycle can be applied effectively to teach science text. High school students were randomly assigned to read either a traditional text passage or a learning-cycle passage. The students in the learning cycle group earned higher scores on concepts comprehension questions at all reading levels.

In addition to Piaget, accommodation and assimilation are also components of constructivism as defined by Fosnot (1989). During concept introduction, students may encounter realities than contradict their existing ideas. Cognitive conflict arises through group dynamics and social exchange as the learner realizes that there may be a contradiction between his or her understanding and what he or she is experiencing (Applefield et al., 2000–2001). Conceptual change theories of instruction are based on constructivist perspectives, and from this view, learning involves interactions between new and existing conceptions. Teaching is more than providing one correct view.

Conceptual change methods which include techniques such as learning cycles and students' changing conceptions have been shown to foster positive student attitudes. Heide (1998) demonstrated that students demonstrated more positive attitudes about science and implemented higher-order thinking skills as a result of constructivist-based conceptual change teaching. Everyone who has ever taught science understands how important for students to believe that they can "like" science and become proficient in it.

Constructivism states that conceptual change is the key to cognitive growth and development, and so conceptual change should become the goal for every good teacher's instruction (Applefield et al., 2000–2001). There is evidence that conceptual understanding of content is higher when students are taught in constructivist classrooms. Current research supports the advantage of conceptual learning over memorization. Constructivism has been very successful in mathematics instruction where students have historically done poorly in terms of understanding certain mathematical concepts, such as giving students relevant examples to solving analogous problems that have some connection to similar problems and prior knowledge (Chen, 1999).

Specifically, Chen (1999), conducted research concerning children's learning and transfer to determine the conditions under which and the extent to which children apply problem solutions from source to target (transfer) problems. Seventy-one children ranging in ages from 8 to 11 years old were recruited from a midsize city. Results showed that children who a learned a general schema (concept) that applied to a problem, had no difficulty answering problems that included formulae and enhanced their flexibility in solving the target problem. In contrast, children in the invariant group who did not learn the concept behind the formula, tended to be tied to the specific formula and so when asked to solve a problem requiring a different formula, they experienced difficulty solving the problems.

Of course, teacher competence can either enhance or sabotage constructivist learning experiences. Success with constructivism is dependent partly on teachers possessing sophisticated epistemologies and being properly trained in the technique. Some researchers go so far as to call traditional teaching techniques "naïve" and constructivism "sophisticated epistemology" (Howard, McGee, Swarty, & Purcell, 2000).

Teachers themselves must embrace constructivist practices during professional development. Berger (1999) showed that teachers must be given learning experiences based on the same pedagogical principles as the ones they are expected to implement with students, and that if teachers are going to teach for understanding, the teachers need to be challenged at their own level of mathematics competence. During a constructivist teacher workshop developed to enhance mathematics instruction, teachers were taught that conceptual learning proceeds to the development of structures, or big ideas that can generalize across experiences. Forty-eight teachers from around the state of Florida were chosen to participate based on geographic location and teaching assignment. The intent was to do a model of K-12 team approach that would later be replicated in each of the six regions of the state. As a result of the teacher education, students scored higher on algebra tests after focusing on the concepts. More important to the students, inquiry learning, which was employed in this study, showed to result in gains over traditional teaching methods in a wide range of students, especially with disadvantaged students deriving greater benefits.

There is evidence that conceptual understanding of content is higher when students are taught in constructivist classrooms. Current research supports the advantage of conceptual learning over memorization. Constructivism has been very successful in mathematics instruction where students have historically done poorly in terms of understanding certain mathematical concepts, such as giving students relevant examples to solving analogous problems that have some connection to similar problems and prior knowledge (Chen, 1999).

Schema Theory

The concept of new information being fitted into a knowledge paradigm that is already there is called *schema*. "A schema is a general knowledge framework that a person has about a particular topic. A schema organizes and guides perception" (Hyde, 1996, p. 58). It is due to this schema concept that true higher level comprehension can occur, not just memorization. When everything connects in the mind, memorization does not have to be relied upon as the core mode of learning. Regardless of the level of sophistication of a student's existing schema, each student existing schema, or knowledge structure, will have a profound impact on what is learned and whether or not real learning (as defined as conceptual change) occurs (Applefield et al., 2002–2001).

Schema is also about putting things into their proper context. Environments where children can interact with their peers, teachers, toys, or instructional materials, enhance their development and their desire to learn. When children play, they use their senses to experience the world; they feel, see, hear, and sometimes taste the world and the objects with which they are playing. When learning is dynamic as in this scenario, new information is placed into its proper schema or context, depending on the situation. Children develop nuances and subtleties otherwise not noticed. In this way, research has shown that too much teacher-directed instruction has either negative effects at worst or neutral effects at best on children's development (Meade, 1999). What is learned tends to be context-bound and tied into the situation in which it is learned (Lave & Wenger, 1991).

Athey defines schema as forms of thought. Athey worked with Meade on a project observing children's actions in relation to schema learning and brain development (Meade, 1999). Meade studied the effects of curriculum intervention on the richness and amount of stimulation teachers give 4-year-olds when they observe children who are fascinated by schemas. The researchers observed 20 nursery school children at play with particular schemas, described as lines, curves, and space order. Meade was interested in the study in terms of brain development and neural pathways. Results showed that the strengthening of neural pathways is enhanced by focused play, a self-organized focus on the schemas, even though adults may not see the play as beneficial. If a child showed interest in a "schema," such as being fascinated by horizontal lines that connect A to B, the teachers would give the students materials for them to connect: ribbon, string, and so forth. They did not give lessons, but simply observed the children. This "provision of diverse experiences" resulted in higher IQ scores because of the enrichment of the childrens' experiences.

As a teacher, I have heard the argument from students many times concerning the "meaningfulness" of the subject matter and topic in any given class. "How will this be important to me in the future?" they ask. Yarlas (1999) argues that interest in a particular subject or class in large part depends on the usefulness and comprehensibility of the information, its meaningfulness to the student, and its ability to be processed and incorporated into a person's existing schema or knowledge structure. Thus, the degree to which information attainment leads to schema enhancement seem to be related directly to the student's interest level for that information. Of course, a student has to take ownership of the subject matter in order to do this. But a good science teacher can take relevant problems in the world today, like global warming or the spread of disease, and make it come alive in class by connecting world news to relevant classroom activities.

Much has been written lately concerning gender and interest in science as a content area. Yarlas (1999) chose physics classrooms and studied the effect gender had on cognitive interest. This was accomplished through assessing a learner's current state of knowledge in a domain, and creating materials that optimized the student's degree of schema enhancement. Students were read passages that contained information about either an expected or an unexpected outcome. Students were asked to either explain or describe information related to these outcomes. Schema enhancement was related to unexpected outcomes, thereby increasing interest. The data strongly supported the prediction that the more interesting the passage, the more learning occurred. Individual interest and gender were covariates because males naturally have more experience with physics and science in general, providing further evidence that supported the central hypothesis of the knowledge-schema theory; that learning increases interest for information in classroom situations where concepts are taught in ways that maximize interest.

Participants in Yarlas's (1999) study demonstrated greater learning for concepts that were related to their own knowledge-schema, than for concepts less related to their schema. This supported the prediction that the more relevant the new information is to existing information already in the child's brain, the more interesting it was for the child, possibly explaining why girls fair poorly in advanced physics classes. Girls start out strongly in science in elementary school, but by middle school drop out in high numbers, thus eliminating any science schema structure on which to build. More about this will be discussed in a later chapter.

Walker (1999) conducted a differential item functional analysis to determine if seventh and eighth grade students participating in the Third International Mathematics and Science Study who were taught mathematics in a constructivist classroom had a higher probability of obtaining the correct answer to mathematics items that measured conceptual, rather than procedural understanding, than students taught in a traditional classroom. Results showed that the constructivist taught students had a higher probability of answering mathematical items that measured conceptual understanding correctly, than students taught in traditional classrooms.

Bruner (1960) contends outside forces or experiences, in addition to growth and maturation, may propel a child from one stage of development into the next. As a cognitive psychologist, the fundamental assumption of Bruner's work is that humans use mental models to represent reality. These models also can be described as modes of representing knowledge and experience:

1. Enactive—from infancy, this mode corresponds to Piaget's sensorimotor stage. This representation is experience translated into action.
2. Iconic—these representations use visual imagery and develop at age two to three.
3. Symbolic—language and mathematics systems and develop from around 7 years of age.

Bruner (1960) moves into an interactionist position in his theory of learning, encompassing constructivism, and emphasizing the roles of exchange between teacher and learner in the acquisition of knowledge. He developed the notion of "the spiral curriculum" whereby the curriculum should involve the mastery of skills that lead to the mastery of higher level skills throughout a child's academic career. For example, the topic of acceleration can be taught in a simple way in first grade, to a more complex way in middle school, to a very detailed formula driven physics class in high school (Howe & Jones, 1993). The example I gave earlier about girls dropping out of middle school science illustrates how spiraling of science content cannot occur when girls drop out of classes.

According to Bruner (1960), learners construct their own meaning through concept formation, and that the learner selects and transforms information, constructs hypotheses, and makes decisions relying on mental models to do so. In order to operationalize Bruner's theories, teachers must be active problem solvers with expectations for the students to be interactive learners. Process is important to Bruner, therefore science education is the perfect vehicle with which to carry out his ideas.

Bruner's (1960) concept formation serves as a vital ingredient in the constructivist classroom. In a study conducted by Discenna and Howse (1998), 22 pre-service elementary education students enrolled in either a physical science or life science course were instructed by one of the authors at a midsized Midwestern university. The researchers were seeking to enhance pre-service teachers' scientific knowledge by changing their notions of science and their epistemological beliefs of on science learning. The authors were interested specifically in describing beliefs that students bring to the science classroom and to science learning as a meaning-making activity, and how these beliefs in science may differ from beliefs about learning.

Both classes stressed problem solving and guided inquiry activities as the method of teaching science. During 15 weeks, the subjects participated in a guided reflection task. After the course, the journals were coded into five "views" of how science should be taught. The most passive view considered science a body of knowledge or set of facts to be memorized by listening. The more active considered science to be the replicating of work by others. A middle view depicted science as existing in objects and that in order to learn, manipulating these objects to discover the "science" behind them was important. Students' ideas changed in a positive way during the semester in terms of science learning, aiding their concept formation. The authors argue that pre-service teachers need more classes in the inquiry/problem-solving tradition with teacher mentors. When pre-service teachers are trained in schema-theory, they begin to understand that the notion of science making and science learning as a meaning-making enterprise. This is very important in fostering the same traits in students once the teachers reach the classroom (Discenna & Howse, 1998).

Cooperative Learning/Social Learning

John Dewey led the way for progressive education reformers at the turn of the century. Dewey held that education was composed of four main objectives: intellectual, moral, social, and aesthetic development. The development of the whole child became the goal. Although the term "constructivism" is never to be found in Dewey literature, his philosophy is the buttress of the whole constructivist movement.

Dewey was the first philosopher to recognize the social as well as the intellectual aspects of learning. Dewey was a proponent of the "social gospel theory" of religion, which stressed the living of one's life for the greater good of others, and for workers and employers to work towards each other's best interests. He wrote of "education as a social function" whereby teaching consists of "social direction" (Dewey, 1916). Note that the role of teaching according to Dewey is not to lecture and impart knowledge, but to direct student activity to discover his or her own knowledge. The classroom consists of a "social environment" (p. 14).

Indeed, social constructivism, supports cooperative learning. According to Vygotsky, children develop in social or group settings. Instead of working alone, children benefit when the teacher serves as the guide, encouraging students to work in groups to discuss issues and challenges that are rooted in real life situations. Teachers thereby facilitate cognitive growth and learning, as do their peers (Anonymous, 1997).

Cooperative learning is based on the Deweyan notion of social learning. Science is the perfect curriculum area for the employment of cooperative learning, since the very nature of scientific exploration includes social learning between laboratory partners. Much empirical evidence exists suggesting that cooperative learning enhances not only a more thorough mastery of skills, but also social and communication skills as well (Slavin, 1985). In sum, Dewey reformulated the framework for education, by stating that learners make sense of new information by placing it in already existing schema, a basic part of constructivism. He dramatically influenced education, and continues to have great presence in the educational arena.

Cognition is viewed as a collaborative process and constructivist thought provides a theoretical basis for cooperative learning, which points toward the powerful social aspect of learning. Students are exposed to their peers' thought processes and opposing views. Constructivists also make use of cooperative learning tasks in relation to learning and comprehension, as well as peer tutoring. Students learn best in situations where they dialog with each other about problems (Applefield et al., 2000–2001).

Johnson and Johnson (1994) state that the effectiveness of cooperative learning has been confirmed by both demonstration and theoretical research. Achievement is greater when learning situations are structured cooperatively rather than competitively or individualistically; students focus both on increasing their own achievement and that of their groupmates. Cooperative learning experiences promote greater critical thinking skills, more positive attitudes about science, greater collaboration skills, better psychological health, and greater perceptions of the grading system as being fair.

Students' notions about science also are affected by cooperative learning. Science is learned by doing and is interwoven with problem-solving activities aimed at involving students in the concepts of science, as well as the pursuit of the scientific method. Teachers have the power to incorporate cooperative learning into their classrooms. According to Yager (1997), principal investigator for the Salish Project, teachers who hold student-centered beliefs were likely to have completed teacher-education programs in which they participated in cooperative learning themselves.

A number of positive outcomes have been attributed to cooperative groups, especially among girls, which will be discussed in more depth in chapter 7. When done correctly, cooperative learning is designed to reduce competitiveness while increasing cooperative spirit, heterogeneous and racial relations, and boosting academic achievement. Teachers must be aware of potential problems with cross-gender cooperative groups because boys can tend to become dominant in the group and suppress the girls' learning (Bailey, 1992).

In a study done by Slavin (1985), 504 mathematics students in Grades 3, 4, and 5 in a suburban Maryland school district were assigned randomly to one of three conditions: team-assisted individualization (cooperative groups), individualized instruction, or without student teams, or control (this group used traditional methods). These treatments were implemented for 8 weeks in Spring, 1981 to evaluate the effects of cooperative learning on achievement, attitudes and behaviors of the students. The cooperative groups gained significantly in achievement than the control group. The results on the "liking of math" scale showed indicated a significant overall treatment effect. Statistically significant overall treatment effects were found for all four of the behavioral rating scales. Six more experiments were conducted. In each of these, classes using cooperative groups were compared to untreated control classes on a variety of dependent measures. In five of the six studies, achievement in the cooperative classes was significantly higher than in the control classes.

Heide (1998) has demonstrated that students' attitudes towards science are more positive when they engage in behaviors such as choosing problems and finding solutions to those problems (student-centered), working in large and small cooperative groups, performing hands-on science laboratory experiments, and learning through conceptual understanding rather than memorization.

Alternative Assessment

As was mentioned in chapter 1, assessment should match instruction. When teachers teach mostly knowledge level fact-based curricula, they assess this way also. The problem lies in how to properly assess students who are learning at higher levels in more constructivist based classrooms. There is no way around the fact that higher level learning cannot be properly assessed by lower-level assessments, yet that is what we do when we assess the standards using high-stakes multiple-choice tests. Among the most important aspects of teaching is reaching agreement on how to determine if the learner can demonstrate in some fashion the desired learning outcome or performance (Applefield et al., 2000–2001).

There are several ways to operationalize ideas about teaching at higher levels. The first is to employ Benjamin Bloom's taxonomy of cognitive levels. In 1956, Bloom developed a brilliant classification system that still stands up today, whereby intellectual behavior important to learning was separated into three domains: cognitive, psychomotor, and affective. The cognitive domain was further divided into six levels, which demonstrate different intellectual skills. These go from the lowest levels of learning to the highest. (Verb examples are included that represent measurable intellectual activity: these ingeniously are used to write daily objectives for lesson plans, since verbs are action, and we can measure [assess] them.)

1. Knowledge: (lowest level) arrange, define, duplicate, label, list, memorize, name, order, recognize, relate, recall, repeat, reproduce.
2. Comprehension: classify, describe, discuss, explain, express, identify, indicate, locate, recognize, report, restate, review, select, translate.
3. Application: apply, choose, demonstrate, dramatize, employ, illustrate, interpret, operate, practice, schedule, sketch, solve, use, write.
4. Analysis: analyze, appraise, calculate, categorize, compare, contrast, criticize, differentiate, discriminate, distinguish, examine, experiment, question, test.
5. Synthesis: arrange, assemble, collect, compose, construct, create, design, develop, formulate, manage, organize, plan, prepare, propose, set up, write.
6. Evaluation: (highest level) appraise, argue, assess, attach, choose compare, defend estimate, judge, predict, rate, core, select, support, value, evaluate (Bloom, 1956, pp. 204–205).

Basically, teachers employing constructivist techniques teach at higher levels than are found in traditional (non-constructivist) classrooms. Lower level instruction is very easy and cheap to evaluate and assess, namely multiple-choice, true/false tests.

Although historically, students taught in both traditional and constructivist classrooms may or may not score similarly on multiple-choice tests, in questions dealing with comprehension, constructivist-taught students had the edge. In a paper presented to the American Educational Research Association conference, it was reported that middle school students who were taught math in a more student-centered conceptual way, had a higher probability of obtaining the correct answer to mathematics items that measured conceptual rather than procedural understanding. The students in this study were 13-year old seventh and eighth graders, who participated in the Third International Mathematics and Science Study (TIMMS). They were administered multiple-choice mathematics items from the TIMMS test as the measure of mathematics ablity. Performance expectations included knowing, using routine procedures, reasoning, and communication. Content areas covered fractions, number sense, algebra, data representation, and analysis and probability. A variant of matrix sampling was used in the test design. Differential item function analysis was used to analyze the data.

Results measured more of a conceptual understanding of mathematics and also a gain for students taught in a more student-centered environment. The students tested also were more successful in obtaining the correct answer to mathematics items that measured conceptual, rather than procedural understanding. According to Walker (1999), students should have acquired a conceptual understanding of the mathematics being taught, knowing not only what to do but *why* they were doing it. The conceptual understanding acquired by these students should enable them to apply their knowledge in new mathematical situations.

In a study that examined teachers' student learning outcome goals and their corresponding assessment practices, Bol and Strage (1996) note that although the national trend is toward integrating the science curriculum into students' daily life aimed at conceptual understanding rather than memory of content, teachers' assessment styles show little correspondence between these goals and actual teaching practices. In fact, teacher developed classroom tests contain mostly low-level questions in terms measuring knowledge. Although teachers' instructional goals were meant to promote higher order thinking skills, the test items included on their assessments do not reinforce those goals. Research also has shown that science teachers stress memorization over conceptual understanding (Gallagher, 1991), thereby reinforcing the need for multiple-choice assessments. The mathematics included in science intimidates many students. Not only do science students have to memorize mathematical formulae, they are then asked to grasp difficult scientific theory in application of the concepts. The problem lies in assessing the higher levels of learning, where memorization is the lowest (Bloom, 1956).

Alternative types of assessment (also called authentic assessment) can be compared and contrasted to more traditional assessment practices that would include standardized tests that feature closed-ended questions. Scores on standardized tests reflect whether or not a student selected the correct answer, but do not reflect the level of comprehension or problem-solving strategies used to arrive at the answer. Bol, Stephenson, and O'Connell (1998) conducted a study where 893 teachers in a large mid-Western urban school district were surveyed to determine assessment practices and their perceptions concerning their practices. Data were analyzed using ANOVAs, and results showed that among teachers in the field, elementary teachers are more likely to use alternative assessment methods than higher grade teachers, and math teachers reported employing alternative assessment more frequently than did science and social studies teachers.

According to Shepard (2000), a broader range of assessment instruments is needed to measure learning goals and processes and to connect assessment directly to ongoing instruction. While multiple choice standardized tests are appropriate for measuring certain levels of acquired knowledge, Shepard suggests more open-ended performance tasks for measuring higher level thinking skills. Not only do teacher made tests measure low-level thinking skills, so do state and district tests. Statewide accountability tests (such as the Standards of Learning in Virginia), used to measure basic knowledge in science, have been corrupted with a heavy-handed rewards and punishment system doled out by administrators who do not reward the excitement of ideas. Types of alternative assessment that would ensure the proper measurement of higher-order thinking would include both informal and formal assessment tools. Some less formal evaluations would include feedback from teacher to student, dynamic ongoing assessment instead of a one-shot final test grade, self-assessment, and teacher assessment. More formal would include portfolios, rubrics, and performance-based assessment. *Assessment for learning must overcome assessment for passing tests.*

According to Gega and Peters (1998), these alternative assessment tools successfully measure higher-order thinking skills:

1. Performance—based assessment-models based on scientific concepts, experiments, journals, written material including papers.
2. Projects—requires self-assessment from start to finish. Students display critical thinking, persistence, inventiveness, and curiosity.
3. Peer or self-designed instruments—rubrics, surveys. Promotes independence and ownership.
4. Interviews—are effective ways of gaining information with students with writing problems or with very early elementary aged students who cannot express themselves in writing.
5. Journals—useful ways to get students to write to learn.
6. Portfolios—a sample of work collected over time, a good self-assessment tool.
7. Concept maps—organizes thoughts and concepts. Helps to see how things are connected, including old and new information.
8. Teacher observations—an informal, on-going tool that puts learning in context.
9. Questioning techniques—open ended questions where there is more than one correct response.

Student-Centered Learning

In discussing the nature of science, Clough (2000) argues that significant consensus exists regarding many issues appropriate for middle and high school students. Some of the most important of these ideas for helping students better understand the nature of science include: science is not the same as technology, a universal, historical scientific method does not exist, science is not completely objective, knowledge is not democratic, words used in science may not mean what students think they do, science is bounded, anomalies do not always result in rejection of an idea, scientific thinking often departs from everyday thinking. Clough suggests that students' understanding is woven into the fabric of their prior experience, which is useful in helping them make sense of new experiences.

The old out-of-date (traditional) trend in middle school science education was the idea that teaching is the transmission of discrete facts, pieces of information and specific processes. The current trend is a broader, more holistic approach that encompasses several areas of instruction, which, in turn, enhances students' understanding and comprehension. Among these are: concepts, processes, applications, attitudes, creativity, and the nature of science. When science instruction focuses solely on transmission of information, only two domains of science are addressed; concepts and processes. Students are presented with a very restricted view of science. This holistic approach develops higher levels of understanding and enables students to "do" science themselves (Daas,

2000). The Western view of the classroom has held that the student is the receiver, not a producer of information. The teacher is idealized as the ultimate source of knowledge and as a highly efficient manager (de Esteban & Penrod, 2000). In a constructivist classroom environment, the teacher's role changes to one of guiding rather than telling the learner the information (Applefield et al., 2000–2001).

For much of the twentieth century, teachers sought to teach facts in a lecture format to students. Now, educators know that teaching children how to think, solve problems and process information is more important than teaching them to memorize facts. Taba adhered to the Deweyan philosophy of education, and agreed with his brand of empiricism (pragmatic instrumentalism) in which "facts" are used to illustrate ideas and not the other way around. Taba believed that teaching should be organized through key concepts, where content should not be seen to dominate any chosen instructional method (Guyver, 1999).

Taba (1932) posits that teachers rely too much on subject matter, forcing them to decide which content to include and exclude by the end of the school year, although she warns against going too far in either direction, stating,

> As a result of a strong reaction against the emphasis on subject matter found in the traditional type of school, progressive education has regarded the child too much as a psychological phenomenon, failing to realize fully that the experience of the child is a product of its contact with the objective materials of its environment. Instead of subject matter alone doing it, the child only is now dictating educational procedure. (pp. 251–252)

Taba's response to increasing the knowledge base is to emphasize the "acquisition, understanding, and use of ideas and concepts rather than facts alone." This reduces the amount of detail to be covered in class, and it provides better conceptual links between pieces of factual information. Broader categories of knowledge like concepts, generalizations and conclusions act to impose structure on factual bits of information, linking these specific bits in categories so that a large amount of specific detail is subsumed within a limited number of ideas. Of course, in our age of standards, teachers have less leeway in this area, and have to teach certain concepts according to the state standards.

Taba (1932) developed a model to categorize information. It is a multipurpose approach that provides an occasional teaching option. The method involved three stages:

1. Students make an exhaustive list of observations, ideas, or concepts.
2. Students gather all similar items together.
3. Students name each category. They then are assigned to category groups and proceed to research their topic. The teacher's role is to facilitate acquisition of relevant information sources. The final product is a report, portfolio, project, or video presentation (Armstrong, 1998).

Taba (1966) also writes that conventional instruction does not reach those adolescents with cultural and educational deficits and that traditional instruction does not meet the needs of these students, because it is incompatible with the needs of those students. Unfortunately, most students today who have educational deficits are poor, minority, and urban, due in part to lower teacher expectations, lack of educational support at home, and lack of funding for poorer schools. Socioeconomic status is the best predictor of both grades and test scores (Bailey, 1992; Wilson, 1997).

Student-centered classrooms have been the topic of empirical study as well. In a study conducted by Chang (1994), constructivist, student-centered classrooms produced students who scored much higher when asked to explain certain scientific phenomena, than students in traditional classrooms. A sample of 363 eighth grade students in a junior high school in Taipei, Taiwan were divided into three groups. All groups were given multiple choice tests, and the scores on the tests were similar in both groups. The difference showed up in the comprehension (as evidenced by explanations) of the subject matter. Also, a teacher main effect appeared in the results of the $3 \times 2 \times 2$ and 3×3 ANOVA analysis, indicating that teachers made significant differences on students' posttest scores. However, results indicate that teacher characteristics, more so than teaching technique, contributed to the results.

Dunkhase, Hand, Shymansky, and Yore (1997) conducted a study comparing a more student-centered environment against a teacher-centered environment and outcomes concerning student attitudes and perceptions. The study focused on student perceptions of their science instruction and student attitudes toward science learning as a function of their exposure to interactive-constructivist teaching strategies aimed at student ideas, utilization of literature integration, and incorporating parents as partners. Among the components of the student-centered environment were interactive-constructivist teaching strategies designed to focus on student ideas, shared control, listening to students' ideas, and making ideas and practices meaningful at the individual student level. Two groups were designated: students from classsrooms where teachers were instructed in constructivist philosophies, and students from classrooms without such instruction. The results showed that attitudes and perceptions were higher in the constructivist/student-centered classrooms than in the traditional classrooms. Of note is the fact that girls experienced the highest rise in attitudes and perceptions concerning the teacher delivery approach, while boys experienced a rise in positive attitudes concerning content.

Active learning includes student participation. Participation encourages students to exchange ideas and viewpoints freely in order to clarify, evaluate, and reconstruct existing schema. In fact, the very effectiveness of a constructivist approach depends on students actively participating in classroom activities. Research shows that constructivist classrooms can increase students' ability to reconstruct their knowledge and that students in constructivist classrooms are challenged to be more active learners (Applefield, 2000–2001; Thomasini & Others, 1990).

Research also shows that students learn more when they have some ownership in the learning process; the basis of constructivism. Yager (1997) states that science students viewed science as more relevant to their daily lives than mathematics was to mathematics students, and that new teachers recently graduated from teacher colleges saw themselves and their classrooms as very student-centered. This study also states that more teachers think they are student-centered when actually their classes are teacher centered, however, students who behaved in student-centered ways were taught by new teachers who held coherent student-centered philosophies of teaching.

Yager et al. (1997) also showed that teacher education programs are crucial for teachers who want to be student-centered in philosophy. Among the findings in this area are these:

- Student-centered actions were not observed in classes taught by new teachers whose philosophies of teaching were not coherent with their practices.

- Students who behaved in student-centered ways were taught by new teachers who held a coherent student-centered philosophy of learning.

- New teachers holding student-centered beliefs were likely to have completed teacher preparation programs were they engaged in cooperative learning, were assessed of their performance in the field, and had strong, close personal relationships with faculty.

- Student-centered teachers were more likely to have completed a longer student teaching experience.

Developmental Stages/Readiness

Jean Piaget made huge contributions towards our current knowledge of intellectual and cognitive development. Piaget brought to light the constructivist notion of readiness; or how children learn in relation to what stage of development they are in currently. Constructivism states that children bring different levels of abstraction, knowledge and understanding to every learning experience, based on cognitive readiness. This concept is where the child-centered constructivism component developed.

The Victorian notion of children as miniature adults was smashed in part by Piaget and his radical views on child development. Piaget's stages of cognitive development are listed below:

1. Sensory motor—birth to 2 years. Babies learn to use their bodies and movement to connect with the outside world. The baby begins to understand how one thing can affect another, and simple notions about time and space.
2. Preoperational—2 to 7 years. Egocentric stage, babies can only consider their point of view. "De-centering" occurs during this stage, where the child comes to realize

that there are other people in the world and that the world does not revolve around them. Not yet logical thinking ... children can fear going down the drain in the bathtub or falling through a crack in the walkway of a fishing pier.
3. Concrete operational—7 to 11 years. Thoughts become more rational and operational. Thoughts are formalized around concrete objects in the presence of the child.
4. Formal operations—11 to 16 years. Thoughts are formalized through hypothetical processes, objects do not have to be present. Abstractions are possible.

One can see the influences of Dewey (1916) in these stages of cognitive development. Dewey states:

> The aim of natural development translates into the aim of respect for physical mobility ... (and that) we find natural development as an aim, enables him to point the means of correcting many evils in current practices, and to indicate a number of desirable specific aims. (p. 115)

Constructivist Learning Environment

Since constructivism does not tell the teacher what former experiences students should have, it does caution teachers against instructional techniques that may limit student understanding. Knowledge is not objective, but the teacher organizes information around conceptual clusters of problems, questions and discrepant situations in order to engage the student's interest (Hanley, 1994).

Driver (1989) has identified certain features that should be present when science is taught from Constructivist pedagogy:

1. Identify and build on the knowledge that learners bring to the lesson.
2. Allow the learners to develop and restructure this knowledge through experiences, discussions, and the teacher's help.
3. Enable pupils to construct for themselves and to use appropriate science concepts.
4. Encourage pupils to take responsibility for their own learning.
5. Help pupils develop understanding of the nature of scientific knowledge, including how the claims of science are validated and how these may change over time (pp. 83–106).

Brooks and Brooks (1993) pose the following as their description of a constructivist classroom setting:

1. They free students from the boredom of fact-driven curriculums and allow focus on large ideas.
2. They turn over to the students the power to follow trails of interest, to make connections, to reformulate ideas, and to reach unique conclusions.

3. They share with students the important message that the world is a complex place in which multiple perspectives exist, and truth is often a matter of interpretation.
4. They acknowledge that learning, and the process of assessment, are elusive and messy endeavors that are not easily managed (p. 32).

To date, many researchers have proposed models of ideal classroom environments. Excellent science classrooms are managed by teachers who use strategies that facilitate sustained student engagement, increase student understanding and comprehension of concepts and scientific knowledge, and encourage student participation in an active learning environment. A recent study suggests that there are advantages of participatory classroom environments, where students construct their own sensory input and make inferences in that information to draw conclusions (Strage & Bol, 1996).

Constructivist classrooms foster communication between students and teachers and among students themselves. Communication apprehension (CA) or "fear to communicate" was studied as a response to teacher philosophies in the classroom. A purposefully selected sample of 61 student teachers during their education program were given the personal report on CA to identify their levels of CA. A Pearson r was used to analyze the data. Results showed that high levels of CA are related to nonconstructivist previous school experience. These people are assumed to have had experienced more traditional teaching styles by their teachers while in school (de Esteban & Penrod, 2000).

Howe and Jones (1993) offer this outline of the major contributors to the constructivist movement and their implication for the science classroom (see Table 1.1). Bowers (1991) offers the following instructional model operationalizing constructivism (see Table 1.2).

DEVELOPMENT OF DIRECT INSTRUCTION

Direct Instruction

Traditional instructional technique is the current instructional strategy based on this philosophy and is based on 100 years of research. The term "direct instruction" was coined by Engelmann. It is the pedagogy currently driving the standardized test movement. From 1966 to 1969, Engelmann was involved in a number of grant-funded projects aimed at exploring the extent to which special instructional methods and innovative curricular approaches would enhance the learning of children. It was during this period that Engelmann coined the term "direct instruction" and formalized the logic and methods for the operationalization of this instructional method. Engelmann's early work focused on beginning reading, language, and math. It was published by Science Research Associates in 1968 under the trade name DISTAR (Direct Instruction System for Teaching and Remediation). Over the past 3 decades, the original curricula have been revised and new ones developed. These curricula have been incorporated into the comprehensive school reform model known as the direct instruction model, which has been implemented in some 150 schools nationwide (Parsons & Polson, 1931).

TABLE 1.1: Constructivism in Science Teaching

SCHOLAR	MAJOR IDEAS OR THEMES
Piaget	Children acquire knowledge by acting and thinking. Knowledge is classified as physical, logico-mathimatical, or social. Development of logical thinking is a maturational process. Understanding of natural phenomena depends on logical thinking ability.
Bruner	Children learn by discovering their own solutions to open-ended problems. Knowledge is represented in enactive, iconic, and symbolic modes. Appropriate ways can be found to introduce children to any topic at any age. The process of learning is more important than the product.
Vygotsky	Children learn through interaction with peers and adults. Knowledge is built as a result of both biological and social forces. Language is a crucial factor in thinking and learning. Children need tasks just above their current level of competence.
Kohlberg	Children learn moral and ethical behavior by example rather than by teaching. Moral development is a slow, maturational process. Moral dilemmas that have no easy solution are part of life. Learning is domain-specific. Misconceptions about natural phenomena interfere with new learning. Both procedural and declarative knowledge are important.
	IMPLICATIONS FOR SCIENCE TEACHING
Piaget	Provide environment to encourage independent action and thought. Distinguish between kinds of knowledge in planning instruction. Be aware of children's level of thinking.
Bruner	Use open-ended problems in science regularly and often. Use all three models of teaching and testing for understanding. Emphasize processes of science. Teach concepts and processes that will lead to further learning.
Vygotsky	Encourage pupils to work together and to learn from each other. Encourage children to explain what they are doing and thinking in science. Set tasks that challenge children to go beyond present accomplishment.
Papert	Make sure that children understand the meaning of their class activities. Make the computer a tool for new learning, not a substitute for a book. Encourage and model thinking about thinking.

Source: Howe and Jones (1993).

TABLE 1.2: Operationalizing Constructivism

	CONCEPT FORMATION	EXPLORATION
Introductory	*Students use descriptive science processes	*Hands on experiences
	DATA INTERPRETATION	CONCEPTUAL
Developmental	*Students identify and investigate relationships	*Student discussion groups
		*Teacher-directed discussions
	*Make inferences	*Students form concepts
	*Use integrated science Processes	*Comparison of student concepts with expert concepts
	Application of principles	Application
Culminating	*Students make predictions and hypotheses	*Concept expansion
	*Support and justify predictions and hypotheses	*Investigation of science/technology/society issues
	*Test predictions and hypotheses	
	*Verify predictions and hypotheses	

Source: Bowers (1991).

There are several working definitions for direct instruction. Direct instruction is described by McDermott (1993) in this way: "Instruction in introductory physics has traditionally been based on the instructor's view of the subject and instructor's perception of the student." The teachers in this scenario are eager to transmit their knowledge to the student. Generalizations often are formulated upon introduction, and students are not actively engaged in the process of abstraction and generalization. The reasoning is almost entirely deductive, very little inductive thinking is involved. McDermott states:

> The trouble with the traditional approach is that it ignores the possibility that the perception of students may be very different from that of the instructor (as is the case in constructivist philosophy). Perhaps most students are not ready or able to learn physics in the way that the subject is usually taught.

In contrast to supporters of constructivism, proponents of direct instruction believe that:

1. External reality does exist independently of the observer.
2. Humans have organized knowledge into systems to better understand reality: such as mathematics, biology, literature, and history among others. The role of teachers is to help students acquire this knowledge.
3. Direct instruction proponents believe that educators are guided by the main concepts of "behavior" and "learning." Behavior is anything students do that is observable. However, direct instruction also cares about how students feel, think, and act.
4. The second main concept is learning, defined as a change in behavior that results in direct interaction with the environment, that is, from teaching-systematic or incidental (Kazloff, LaNunziata, & Cowardin, 1999).

Marchland-Martella, Martella, and Lignugaria-Kraft (1997) developed a system to use while observing practicum teachers delivering a direct instruction lesson (see Table 1.3).

TABLE 1.3: Definitions of Correct Direct Instruction Behaviors

PRESENTATION

Cue—Focus word, phrase, or question (e.g., what word?, get ready) as indicated by program format or as specified by teacher

Pause—At least a 1 second waiting time (preferably 2 seconds)

Signal—Hand, touch, or auditory response presented by teacher which initiates a pupil response

RESPONSES

Group—Two or more pupils respond simultaneously and correctly

Individual—Pupil responds correctly

SIGNAL ERROR CORRECTIONS

Address—Corrects within 3 seconds after group error occurs; addressed to group; positive tone (without negative comments or gestures); tells group what they have to do (e.g., *I've got to hear everyone. You have to wait until I signal*)

Repeat—Repeat original presentation to test group's response; positive tone (without negative comments or gestures)

RESPONSE ERROR CORRECTIONS

Model—Corrects error within 3 seconds after group/individual error occurs; addresses model to group (if group response) or individual (if individual response); positive tone (without negative comments or gestures); demonstrates correct response to pupil(s)

Test—Requests group/individual to respond again using original cue provided before error occurred; addresses test to individual if individual response or group if group response; positive tone (without negative comments or gestures)

PRAISE STATEMENTS

Specific—Precise statement that reflects a positive response to a desired behavior (e.g., *Nice job saying* brother) which is delivered after an appropriate behavioral or academic response (e.g., pupil is sitting quietly with hands folded)

General—Global or broad statement that reflects a positive response to a desired behavior (e.g., Super) which is delivered after an appropriate behavioral or academic response (e.g., student completes homework assignment)

Source: Marchland-Martella et al. (1997).

As is shown, direct instruction relies on teacher-centered lecture, with students having one correct answer. According to McDermott (1993), the reason many students do not understand subjects like physics is that the teachers rely solely on transmitting knowledge from themselves to their students, and that the trouble with the traditional approach to instruction is that it ignores the possibility that the students may have a different perception of the subject than the teacher has. Most science teachers view their students as mini-versions of themselves, when that is not the case.

McDermott (1993) also offers these shortcomings of traditional instruction:

1. Facility in solving quantitative problems is not an adequate criterion for understanding. Questions that require qualitative reasoning and verbal explanation are essential.
2. A coherent conceptual framework is not usually the outcome of traditional instruction: students must participate in the process of constructing qualitative models that can help them understand relationships and differences among concepts.
3. Certain conceptual difficulties are not overcome by traditional instruction. Persistent conceptual difficulties must be addressed by repeated exposures in more than one context.
4. Growth in reasoning ability does not result from traditional instruction and scientific reasoning skills must be cultivated.
5. Connections among concepts, formal representations, and the real world are lacking after traditional instruction. Students need practice in interpreting physics formalism and relating it to the real world.

Teaching by telling is an ineffective mode of instruction for most students. Students must be intellectually active to develop a functional understanding (McDermott, 1993, p. 105).

Another term used for direct instruction is "instructivist" approach, a term coined by Finn and Ravitch in 1996 in their report *Education Reform 1995–1996: A Report from the Educational Excellence Network to its Education Policy Committee and the American People*. Finn and Ravitch argue that constructivism is faddish and that it excludes content. In a paragraph headed "The Romance of 'Natural Learning', " they posit that constructivism is "hostile to standards, assessments and accountability" (p. 106).

Finn and Ravitch (1996) also argue that too much constructivism means kids who can neither read nor write, although they may have curiosity and self-esteem. Although keenly pro instructivist, they also argue for a balance in the classroom. The best teachers are not a slave to dogma, they are able to employ constructivist and instructivist techniques as the situation and child require.

In his book *Cultural Literacy*, Hirsch wrote that a content-based curricula was preferred, which ran counter to progressive educators' beliefs that natural development, process and critical thinking skills were goals to be met by education. For Hirsch (as cited in M. Berube, 1994, pp. 1–11), the fault with American education lay with the theories of Rousseau whose ideas influenced John Dewey, claiming that Dewey advocated the content-neutral curriculum.

Advanced Organizers

Ausubel's contribution to learning theory includes his belief that humans acquire meaningful learning through an interaction of newly learned information with relevant existing ideas in cognitive structure. Ausubel explored the process of what he calls meaningful learning and how it relates to a learner's cognitive structure. His "theory of meaningful verbal learning" was unveiled in his 1963 book *The Psychology of Meaningful Learning*. He also promotes the arrangement of school curriculum to match student readiness, which shows influence of Piaget.

Although Ausubel (1963) openly supports direct instruction, he also writes that the learner must make an intellectual link between newly learned information and that previously stored in his or her cognitive structure. Because of this connection, retention is greater and understanding is significant.

In order to facilitate new learning, Ausubel (1963) advocates advanced organizers; outlines of material yet to be learned, a type of summary of material that highlights key concepts and propositions for the students. Knowing that the brain builds knowledge in a hierarchical structure and by assimilating new knowledge with the help of advanced organizers, the learner builds anchors for future knowledge.

There is empirical support for direct instruction. In a study supporting traditional methods, 138 students (including 23 mildly handicapped students) in Grades 4 through 6 participated in a study aimed at comparing the effectiveness of two teaching techniques (direct instruction versus discovery teaching) in three elementary schools in a suburban Chicago school system on achievement. Students were randomly assigned to one of two treatments: direct instruction or discovery teaching. A 2×5 factorial design was employed. Results showed that students in both groups learned equally well as measured by a posttest. However, students in the discovery treatment group outperformed their direct instruction peers on a delayed posttest administered 2 weeks after the treatment ended (Bay & Others, 1992).

Project Head Start, a grant funded by the U.S. Department of Education between 1969 and 1972, was directed by Englemann. The purpose of the grant was to provide a comparison of the different models of educational programs for disadvantaged children. Children in three Engelmann-Becker models were compared with children in other models of instruction. This was called the largest controlled comparative study of teaching methods in history. The Engelmann-Becker model worked with twenty school districts to implement effective instructional programs in Grades 1 through 3 as part of Head Start. Research focused on specific variables that made a difference in student performance. Results showed that students in direct instruction classrooms had placed first in reading, math, spelling and language. Even though no other model was as effective, Direct Instruction has been spurned by the majority of the educational establishment (Anonymous, 2000).

Direct instruction advocates posit that behavior is anything students do, and therefore, learning is a change in behavior (feeling, thinking, acting) that results from interaction with their environment. The instructivist approach in education means that educators draw on literature on how students learn to design appropriate curricula, and focus on changes in students' behavior (learning) as a way of tracking progress (Applefield, 2000-2001).

Operationalizing Instructivism

According to Kozloff, LaNunziata, and Cowardin (1999), there are basically three distinct approaches to teaching using the instructivist method:

1. *Applied behavior analysis:* (Kozloff et al., 1999; Kozloff, LaNunziata, Cowardin, & Bessellieu, 2000). The first branch of instructivist technique is really a combination of practices derived from years of experimental research on how environmental events and arrangements affect learning and principles of operant learning, found in the work of B. F. Skinner. These tenants are as follows:

    a. Methods for examining the interaction of students with their environments so that relationships may be discovered, that is, one can find out how a student's learning is helped or hurt by such things as difficulty, pacing, and assistance from the teacher, or the nature of their interaction with peers.
    b. Guidelines for using knowledge of functional relationships between environmental features and a student's learning, to design instruction that is consistent with a student's skills.

c. Methods of evaluating the adequacy of curriculum and instruction by tracking students' learning, and revising curriculum and instruction accordingly.

2. *Precision teaching:* Developed by Ogden Lindsey and associates. Lindsey based precision teaching on Skinner's discovery that the rate of behavior (# of occurrences/time) is a dimension of behavior, and not just a measure of the behavior. This implies a difference in fluent versus nonfluent behavior. The following are features:

   a. Teachers identify and teach the "tool skills" (component or elemental skills and knowledge) needed to learn complex skills and knowledge. For example, listening to a teacher, taking notes, having fluency with math facts, etc. When students are not fluent with tool skills (reading and writing), they are not able to learn complex skills.
   b. Teachers provide carefully planned, short practice sessions on older and new learning to strengthen retention.
   c. As students master component skills, teachers help students to assemble component skills into complex activities.
   d. Teachers help students keep track of their own progress.

3. *Direct instruction:* (Adams & Englemann, 1996). This third branch of the instructivist approach grew out of the work of Englemann and his work with disadvantaged children. Direct instruction was compared with 12 other methods of instruction during the largest educational study ever conducted and results showed that direct instruction was superior in fostering reading and math skills, higher-order cognitive skills, and self-esteem.

   a. Direct Instruction focuses on cognitive learning-concepts, propositions, strategies, and operations.
   b. Curriculum development involves three analyses: knowledge, communication, and student behavior.
   c. Instruction teaches concepts, strategies and operations to greater mastery and generality. Direct instruction focuses on big ideas.
   d. Concepts are not taught in isolation from each other.
   e. The analysis of knowledge is used to create student-teacher communication.
   f. Lessons are arranged logically so that students first learn what is needed to grasp later concepts.
   g. Lessons are formatted so teachers know what to say and what to ask that enables students to reveal understanding and/or difficulties.

    h.   Lessons are followed by independent and small group activity.
    i.   Gradually, instruction moves from teacher guided to more student guided.
    j.   Short proficiency tests are used about every ten lessons (Kozloff et al., 2001).

Bowers (1991) offers a way to differentiate between constructivism and more traditional direct instruction teaching techniques in actual classroom situations. Bowers cites Tickle who writes that the core teaching issue in middle school is the tension between the two instructional approaches; as he puts it, "one emphasizes the mastery of skills in content and the other stresses providing for the developmental needs of young adolescents" (pp. 4–9). Bowers also argues for a non-content-area-specific learning approach that would emphasize the whole child and not just rote memorization.

In differentiating between the two methods of teaching, Bowers (1991) includes examples of behaviors that would occur during each educational experience. Representing constructivism, Bowers has combined the inductive thinking theory set forth by Taba, and the learning cycle, which began several years ago as part of the Science Curriculum Improvement Study. Barman (1989) has modified the terminology of the learning cycle to make it more meaningful for elementary school teachers. Representing the traditional or direct instruction approach, Bowers (1991) cites Ausubel's (1963) "advanced organizer." The following sets of behaviors are grouped as: (1) **Introductory**: the beginning of the daily lesson, (2) **Developmental**: the operationalizing of the lesson, and (3) **Culminating**: the summation of the daily lesson. I have combined inductive thinking and learning cycle behaviors to represent the functions of a constructivist classroom and the advanced organizer for the direct instruction classroom.

Bowers (1991) offers in Table 1.4 the instructional model operationalizing direct instruction.

A paradigm describing traditional versus constructivist classroom environments is provided by de Esteban and Penrod (2000) (See Table 1.5).

In trying to bring all of this research into focus, one needs only to remember that there is a difference in techniques between constructivist teachers and more traditional teachers, and that both constructivist and traditional pedagogies alike, have meaningful and valuable components. Since this book focuses on assessment rather than instruction, it is important nonetheless to view instructional practices with assessment in mind, since we cannot have one without the other. Traditional and progressive teaching techniques must by their very nature result in different assessment strategies.

Progressive education seeks to educate the whole child, hence the use of constructivist strategies that the movement has adopted. It is important for children to *like* school, and to be engaged in relevant, meaningful tasks that force children to think, instead of merely parroting back information. A progressive notion indeed.

TABLE 1.4: Operationalizing Direct Instruction

INTRODUCTORY

*Clarify objectives
*Give examples
*Define context
*Prompt learner's prior knowledge and experience

DEVELOPMENTAL

*Directed teaching
*Organization of tasks
*Logical order of material

CULMINATING

*Students integrate new learning and prior knowledge
*The teacher promotes logical and critical approach to information
*Students resolve conflicting information and misconceptions

Source: Bowers (1991).

TABLE 1.5: Traditional Versus Constructivist Classroom Environment

TRADITIONAL CLASSROOM	CONSTRUCTIVIST CLASSROOM
Curriculum is presented part to whole with emphasis in basic skills. Strict adherence to fixed curriculum is highly valued.	Curriculum is presented whole to part with emphasis on big concepts.
Curricular activities rely heavily on textbooks and notebooks.	Pursuit of student questions is highly valued.
Students are viewed as black slates onto which information is etched by the teacher.	Curricular activities rely heavily on primary sources of data and manipulative materials.
Teachers generally behave in a didactic manner, disseminating information to students.	Students are viewed as thinkers with emerging theories about the world.
Teachers seek the correct answer to validate students' learning.	Teachers generally behave in an interactive manner mediating the environment for the students.

Assessment of student learning is viewed as separate from teaching and occur almost entirely through testing.	Teachers seek the students' points of view in order to understand students' present conceptions for use in subsequent lessons.
Students primarily work alone.	Assessment of student learning is interwoven with teaching and occurs through teacher observations of students at work and through students' exhibitions and portfolios.

Source: de Esteban and Penrod (2000).

REFERENCES

Adams G. L., & Englemann, S. (1996). *Research on direct instruction: 25 years beyond DISTAR.* Seattle, WA: Educational achievement Systems.

Andrews, J. F. (1985, June). *Deaf children's acquisition of prereading skills using the reciprocal teaching procedure.* Paper presented at the Council of American Instructors of the Deaf, Florida School for the Deaf, St. Augustine, Florida.

Applefield, J. M., Huber, R., & Moallen, M. (2000–2001, December-January). Constructivism in theory and practice: Toward a better understanding. *The High School Journal, 84*(2), 34–53.

Armstrong, T. (1998). *Awankening genius in the classroom.* Alexandria, VA: ASCD.

Ausubel, D. (1963). *The psychology of meaningful verbal learning.* New York: Greene & Stratton.

Bailey, S. M. (Ed.). (1992). *How schools shortchange girls: The AAUW Report, a study of major findings on girls and education.* New York: Marlowe.

Bandura, A. (1977). *Social learning theory.* Upper Saddle River, NJ: Prentice Hall.

Barman, C. R. (1989). A procedure for helping prospective elementary teachers integrate the learning cycle into science textbooks. *Journal of Science Teacher Education, 1*(2), 21–26.

Bay, M., & Others. (1992). Science instruction for the mildly handicapped: Direct instruction versus discovery teaching. *Journal of Research in Science Teaching, 29*(6), 555–570.

Berger, S. L. (1999). *Opening the gate: Changing the attitudes and practices of teachers through a constructivist professional development model* Unpublished doctoral dissertation, Florida State University.

Berube, M. R. (1994). *American school reform: Progressive, equity, and excellence movements, 1883–1993.* Westport, CT: Praeger.

Bevevino, M. M., Dengel, J., & Adams, K. (1999). Constructivist theory in the classroom: Internalizing concepts through inquiry learning. *The Clearing House, 72*(5), 274–275.

Bloom, B. S. (Ed.). (1956). *Taxonomy of educational objectives, the classification of educational goals, handbook I: Cognitive domain.* New York: David McKay.

Bol, L., & Strage, A. A. (1996). High school biology: What makes it a challenge for teachers? *Journal of Research in Science Teaching, 33*(7), 125, 753–772.

Bol, L., Stephenson, P. L., & O'Connell, A. A. (1998). Influence of experience, grade level, and subject area on teachers' assessment practices. *The Journal of Educational Research, 91*(6), 323–330.

Bowers, R. S. (1991). Effective models for middle school science instruction. *Middle School Journal, 22*(4), 4–9.

Brooks, J. G., & Brooks, M. G., (1993). *The case for constructivist classrooms*. Alexandria, VA: Association for Supervision & Curriculum Deve.

Bruner, J. (1960). *The process of education*. Cambridge, MA: Harvard University Press.

Chang, M. M. (1994). *Constructivist and objectivist approaches to teaching chemistry concepts to junior high school students*. Paper presented at AERA, New Orleans. LA.

Chen, Z. (1999). Schema induction in children's analogical problem solving. *Journal of Educational Psychology, 4,* 703–715.

Clough, M. P. (2000). The nature of science: understanding how the game of science is played. *The Clearing House, 74*(1), 14.

Daas, P. M. (2000). Preparing coaches for the changing game of science: teaching in multiple domains. *The Clearing House, 74*(1), 39–41.

de Esteban, M., & Penrod, K. (2000). *Effect of teaching philosophy orientation on students' levels of communication apprehension* (Research report No. 143). Brookings, SD: South Dakota St. University, Department of Undergraduate Teacher Educaton, College of Education and Counseling.

Dewey, J. (1916). *Democracy and education*. New York: The Free Press.

Discenna, J., & Howse, M. (1998). *Biology and physics students' beliefs about science and science learing in non-traditional classrooms*. Paper presented at AERA, San Diego, CA.

Driver, R. (1989). The construction of scientific knowledge. In R. Miller (Ed.), *Doing Science: Images of science in science education* (pp. 83–106). London: Falmer Press.

Dunkhase, J. A., Hand, B. M., Shymansky, J. A., & Yore, L. D. (1997). *The effect of a teacher enhanced project designed to promote interactive-constructivist teaching strategies in elementary school science on students' perceptions and attitudes*. Paper presented at the School of Science and Mathematics conference, Milwaukee, WI.

Ebenezer, J. V., & Haggerty, S. M. (1999). *Becoming a secondary school science teacher*. Upper Saddle River, NJ: Prentice Hall.

Finn, C., & Ravitch, D. (1996). *Education Reform 1995-1996: A report from the educational excellent network to its education policy committee and the American people*. Retrieved from www.edexcellence.net/library/epctot.html

Fosnot, C. T. (1989). *Inquiring teachers, inquiring learners: A constructivist approach for teaching*. New York: Teachers College Press.

Gallagher, J. J. (1991). Prospective and practicing secondary school science teachers' knowledge and beliefs about the philosophy of science. *Science Education, 75,* 121–133.

Gega, P. C., & Peters, J. M. (1998). *Science in elementary education*. Upper Saddle River, NJ: Merrill.

Guyver, R. (1999). *National curriculum: Key concepts and curriculum controversy*. Retrieved from www.Harris8.freeserve.co.uk/rguyver.html

Hanley, S. (1994). *On constructivism*. College Park, MD: Maryland Collaborative for Teacher Preparation.

Heide, C. L. (1998). *Attitudes of eighth grade honors students toward the conceptual change methods of teaching science (middle school students*. Unpublished doctoral dissertation, Northern Arizona University.

Howard, B. C., McGee, S., Schwartz, & N., Purcell S. (2000). The experience of constructivism: transforming teacher epistemology. *Journal of Research on Computing in Education, 32*(4), 455.

Howe, A. C., & Jones, L. (1993). *Engaging children in science* (1st ed.). New York: MacMillan.

Hyde, J. S. (1996). *Half the human experience: The psychology of women* (5th ed.) Washington, DC: Heath & Co.

Johnson D. W., & Johnson, R. T. (1994). *Learning together and alone: Cooperative, competitive, and individualistic learning* (4th ed.) Boston: Allyn & Bacon.

Kozloff, M. A., LaNunziata, L., & Cowardin, J. (1999). *Direct instruction in education*. Retrieved 2001 from www.uncwil.edu/people/Kozloffm/diarticle.html

Kozloff, M. A., LaNunziata, L., Cowardin, J., & Bessellieu, F. B. (2001). Direct instruction: Its contribution to high school achievement. *The High School Journal, 84*(2), 54–70.

Lave, J., & Wenger, E. (1991). Authors, text and talk: The internalization of dialogue from social interaction during writing. *Reading Research Quarterly, 29*, 201–231.

Marchland-Martella, N. E., Martell, R. C., & Lingnugaria-Kraft, B. (1997). Observation of direct instruction teaching behavior. Determining a representative instruction teaching for supervision. *International Journal of Special Education, 12*(2), 32.

McDermott, L. C. (1993). How we teach and how students learn. *Annals of the New York Academy of Science, 701*, 9–19, 295.

McNichols, T. J. (2000). Deconstructing constructivism: The Kantian connection. *The Journal of Philosophy and History of Education, 49*, 4.

Meade, A. (1999, November). *Schema learning and its possible links to brain development*. Paper presented at at a seminar at the Children's Hospital of Michigan, Wayne State University.

Musheno, B. V., & Lawson, A. E. (1999). Effects of learning cycle and traditional texts on comprehension of science concepts by students at differeing reasoning levels. *Journal of Research in Science Teaching, 36*(1), 23–37.

Palinscar, A., David, Y., & Brown, A. (1984). Reciprocal teaching of comprehension-fostering and comprehension-monitoring activities. *Cognition and Instruction, 1*(2), 121–122, 145, 153, 164.

Palinscar, A., David, Y., & Brown, A. (1992). Using reciprocal teaching in the classroom: A guide for teachers. *The Brown/Campione Research Group, 12*, 42–47.

Parson, J., & Polson, D. (1931). *Siegfried Engelmann*. Retrieved April 11, 2008, http://psych.athabascau.ca/?html/387/Openmodules/Engelmann/Engelmannbio.shtml

Richardson, V. (1999). Teacher education and the construction of meaning. In G. Griffin (Ed.), *The education of teacgers: Ninety-eight yearbook of the National Society for the Study of Education* (Part 1, pp. 46, 145–166). Chicago: University of Chicago Press.

Shepard, L. A. (2000). The role of assessment in a learning culture. *Educational Researcher, 29*(7), 4–14.

Slavin, R. (1985). An introduction to cooperative learning research. In *Learning to cooperate, cooperating to learn*. New York: Plenum Press.

Strage, A. A., & Bol, L. (1996). High school biology: What makes it a challenge for teachers? *Journal of Research in Science teaching, 33*(7), 753–772.

Taba, H. (1932). *The dynamics of education: A methodology of progressive educational thought.* London: Kegan Paul, Trench, Trubner & Co.

Taba, H. (1966). *Teaching stragegies for the culturally disadvantaged*. Chicago: Rand McNally.

Thomasini, N. G., & Others. (1990). *Teaching strategies and conceptual change: sinking and floating at elemtary level*. Paper presented at AERA, Boston.

Walker, C. (1999). *The effect of different pedagogical approaches on mathematics students achievement*. Paper presented at AERA, Montreal, Canada.

Wilson, W. J. (1997). *When work disappears: The world of the new urban poor*. New York: Vintage Books

Yarlas, A. S. (1999). *Schema modification and enhancement as predictors of interest; at test of the knowledge-schema theory of cognitive interest*. Paper presented at AERA, Montreal, 29–35.

Yager, R. E. (Ed.) (1997). *SALISH: A research project dedicated to improving science and mathematics teacher education. Secondary science and mathematics teacher preparation programs: Influences on new teachers and their students*. Washington, DC: U.S. Department of Education.

READING 2

A CASE FOR CULTURALLY RELEVANT TEACHING IN SCIENCE EDUCATION AND LESSONS LEARNED FOR TEACHER EDUCATION

By Felicia Moore Mensah

In this article, the researcher discusses three elementary pre-service teachers' experiences in co-planning and co-teaching a Pollution Unit in a 4th–5th grade science classroom in New York City. The study makes use of microteaching papers, lesson plans, researcher classroom observations, interviews, and informal conversations to elicit lessons learned from implementing culturally relevant teaching in science education. Examples of pre-service teachers' planning process, culturally relevant teaching examples, assessment of student learning, and reflections on microteaching are also presented as exemplars and interpretations of culturally relevant science teaching. Implications from the study are discussed in terms of support to enact culturally relevant teaching in urban elementary classrooms and in pre-service science teacher education.

Keywords: culturally relevant teaching, science education, teacher education, multicultural education

The literature on multicultural teacher education emphasizes the preparation of teachers for diverse classrooms, with a great deal of the literature focusing on the preparation of White teachers for communities that have been traditionally underserved. In a literature review by Sleeter (2001), she discussed the effects of various pre-service teacher education strategies, ranging from recruiting and selecting students to program restructuring. She concluded that most of the research addresses teacher attitudes and the lack of knowledge that pre-service teachers have about multicultural teaching. She also acknowledged that research has not addressed how to populate the teaching profession with excellent multicultural and culturally responsive teachers.

To take this further, Furman (2008) conducted a review of the literature of multicultural teacher education (MTE) by examining previous published reviews and synthesized the field over the past two decades. Specifically, he "examine[d] the ways in which the problem of MTE is established and understood, how these issues have been approached by various scholars, and the evolution and current state of the field" (p. 57). He noted specific contributions, limitations, and tensions faced by multicultural researchers and the field of MTE, and then discussed two major tensions within the field—the demographic tension: how best to prepare teacher candidates for increasingly diverse schools, and the effectiveness tension. The effectiveness tension is the one that connects to this current study.

Furman (2008) argued that the effectiveness tension for MTE lies in teacher education. He stated that "teacher education itself must be culturally responsive" (p. 69), yet in his examination of research reviews on teacher education programs, it was shown that this was not the case. Similarly, Villegas and Lucas (2002) argued that in order to move the field of teacher education, and equally, MTE, a vision of teaching and learning in a diverse society is needed. Furthermore, this vision should be used to systematically guide the infusion of multicultural issues throughout the pre-service curriculum. The next logical step is culturally relevant teaching (CRT). In other words, how might teacher education address the effectiveness tension? What is a framework of culturally relevant teaching that can be used as both a curriculum in the preparation of pre-service teachers and a strategy for teaching an increasingly diverse public school populace? Teacher education is one context to promote culturally relevant teaching practices with the hope that these kinds of practices will be implemented in classrooms. This current study argues for the teaching and learning of CRT principles in science teacher education as a means of preparing all teachers for diverse classrooms.

Ladson-Billings (1995a) explained that culturally relevant teaching rests on three criteria or propositions: "(a) students must experience academic success; (b) students must develop and/or maintain cultural competence; and (c) students must develop a critical consciousness through which they challenge the status quo of the current social order" (p. 160). In order for these propositions to be accomplished for student learning, we cannot assume that they will organically become the practices of teachers. In fact, teaching in culturally relevant and or multicultural ways requires a knowledge base about teaching for cultural diversity (Gay, 2002) as well as understanding the sociopolitical context of multicultural education (Nieto, 2004). Ladson-Billings (1995b) argued that "not only must teachers encourage academic success and cultural competence they must help students to recognize, understand, and critique current social inequities" (p. 479). Culturally relevant teaching practices give students the opportunity to learn in ways that are affirming, validating, and connected to their interests and backgrounds (Ladson-Billings, 1994). In order to have classrooms that are culturally relevant, we have to make stronger connections to the lives of students.

Therefore, in order to consider practices of culturally relevant teaching for urban school learners, a more foundational level is to promote culturally relevant teaching at the pre-service teacher level. This view encourages pre-service teachers to learn and implement these principles into practice. Ideally, if pre-service teachers understand CRT principles in teacher education, they can begin to transform theory into practice, and use these principles within the contexts of authentic classroom settings. The ultimate goal is to have these principles become an elemental part of their regular teaching practices as early career teachers and beyond.

Ladson-Billings in her initial interpretation of culturally relevant teaching did not focus on science. Content specific application of culturally relevant teaching seems to be the missing link in the research literature on CRT and teacher education. It has been shown that science is not a priority subject in elementary school programs (Spillane et al., 2001; Tate, 2001). This current study argues for the teaching and learning of CRT principles in science teacher education as a means of preparing all teachers and for the application of CRT in elementary science classrooms. The study in particular focuses on three elementary pre-service teachers' experiences in planning, teaching, and assessing a Pollution Unit in a 4th–5th grade science classroom and how they incorporate principles of culturally relevant science teaching. The study examines what the process was like for learning and enacting culturally relevant teaching in science for these three teachers, and how successful they were in promoting academic success, cultural competence, and critical consciousness for them as teacher learners in a science methods course and for urban students in an elementary science classroom. The research questions that guided this study were: what supports are needed in the preparation of pre-service teachers who focus on planning, teaching, and assessing science lessons and teaching in culturally relevant ways? What lessons are learned in preparing pre-service teachers to incorporate culturally relevant teaching in urban elementary science classrooms and their learning to become culturally relevant science teachers?

DESCRIPTION OF RESEARCH SITES

University Methods Course

This article was written with data collected from an elementary science methods course at a large urban university located in New York City. Pre-service teachers (PSTs) were enrolled as graduate students seeking initial teacher certification, and the science methods course was one course within their teacher education program. The methods course lasted 16 weeks and focused on "science teaching through inquiry, multiculturalism, social justice, and the relevance of science to everyday life in the city and to urban school students" (Gunning & Mensah, 2010, p. 176). Out of the class of 14 pre-service teachers, three were asked to participate in this study: Niyah, Hope, and Estelle. All names used in the study are pseudonyms. The three PSTs were of diverse racial/ethnic background and language (Table 2.1). Niyah, Hope, and Estelle were interested in teaching the 4th grade. They had some teaching experiences or had worked with elementary students, yet they had no prior science teaching experiences. Because they did not have student teaching placements during the semester they were enrolled in the methods course, they were placed together in an East Harlem elementary school to conduct classroom observations (approximately 1–2 hours per week over the semester) and to support one another in planning, teaching, and assessing their science microteaching unit. As the instructor of the course, I observed them teaching their lessons. At the conclusion of the course, I invited them to participate in this study, which meant collection and analysis of their course assignments, observations and debriefings in the field, and interviews.

TABLE 2.1: Teacher Profiles

PSTs (PSEUDONYM)	RACE/ETHNICITY*-LANGUAGE	GRADE-LEVEL INTEREST	TEACHING EXPERIENCE
Niyah	Asian American—does not speak Spanish	4	Private tutoring and math teacher in an afterschool academy
Hope	White—does not speak Spanish	4	Afterschool programs
Estelle	Latina/African American-speaks Spanish fluently	3–4	Part-time instructor at Kaplan's K-12 Learning Services

Note. PSTs = pre-service teachers; *self-identified.

John David Elementary School

John David Elementary School (JDES) is a K-5 urban elementary school located in East Harlem, New York City. From the Department of Education school profile the student enrollment, at the time of this study, was 156 students (57% African American; 41% Latino/a; 3% Asian and other; 55% boys and 45% girls). The student body included 6% English Language Learners (ELL) and 26% special education students (http://schools.nyc.gov/Accountability/tools/report/default.htm). The school was 60% eligible for Title I funding. JDES offered extracurricular programs that students attend during the school day, such as swimming, music and field trips.

In past years, the science program has been strained. For example, in 2004, the science teacher resigned. From that time, science was taught by substitute teachers who had no background in science teaching or teaching. I began working with the school more consistently and in greater capacity in spring 2005 and served as the science teacher for one semester on a voluntary basis. In 2006, they hired a science teacher, Miss Phoebe, who had no previous teaching experience and no content background in science. Since her appointment, I have been working with Miss Phoebe as the science specialist and assisting in developing the science program and her professional development as the K-5 science teacher. The school has been supportive in nurturing a university-school community partnership (Mensah, 2011). PSTs each semester complete observations, microteaching, and work with the students in the science classroom, and doctoral students have done research in the school. In addition, the regular classroom teachers at all levels have allowed PSTs to make observations in their classrooms and to teach integrated science lessons as part of the PSTs' microteaching assignment.

METHODS

Sources of Data

The study makes use of course artifacts from the elementary science methods course, such as the PSTs' pre-assessment interviews with elementary learners, microteaching papers, and lesson plans as the primary source. Individually, the PSTs conducted pre-assessment interviews with one to two elementary students about the topic of pollution. They collaborated and used the interview data to plan culturally relevant lessons and assessments that would promote critical thinking around the topic of pollution. The format of the microteaching paper covered topics such as student misconceptions from conducting the pre-assessment, reflections on their microteaching, and professional growth as a science teacher. Other forms of data consisted of the PSTs' surveys (initial, to gather demographic data, and final, to gather comments about their learning and the course), and interviews (40–50 minutes, conducted at the end of the semester). The interview questions asked the PSTs about teaching in diverse classrooms, planning science, teaching science, future plans in teaching, and professional growth. Finally, a research journal was kept containing observations of the PSTs' microteaching at JDES, informal (debriefing) conversations after their microteaching lessons, and notes from the course.

Analysis of Data

The multiple data sources were coded using methods of constructivist grounded theory and comparative analysis for emergent themes (Charmaz, 2006; Strauss & Corbin, 1989). Using the microteaching papers as the primary data source, several codes and categories were generated from each of the three microteaching papers. These codes and categories were compared across the three papers individually and then across them collectively. Specific attention was given to culturally relevant connections on the topic of pollution. Categories were collapsed into larger themes, such as the planning process, teaching the lessons, assessing student learning, and reflecting on professional growth. These were also the major headings of their microteaching papers. The additional data sources were triangulated to corroborate evidence from the microteaching papers in order to shed light on themes generated from the study (Creswell, 2007). For example, the lesson plans that each of the three PSTs submitted—two lessons per teacher to create their Pollution Unit—were viewed and used to enhance descriptions from the microteaching papers and the classroom observations of their microteaching. The PSTs had to collect, assess, and discuss student learning from their individual lesson plans as part of their microteaching paper. The themes generated from the data analysis were enhanced through member checks and peer debriefing (Guba & Lincoln, 1989). The themes, after a brief description of the Pollution Unit, are given as emergent assertions in the preparation of teachers for culturally relevant science teaching.

FINDINGS

Background of Lesson

The findings and discussion are written with a focus on pre-service teacher education and lessons learned from teaching in culturally relevant ways in an elementary science classroom and using CRT principles in science teacher education. Niyah, Hope, and Estelle taught a six-day Pollution Unit for one 4th–5th grade classroom. They took turns as the lead teacher for 2 days, with the other two acting as support in the classroom. The "team teaching with co-role support" model (Estelle, Microteaching paper) served to be very beneficial for science teaching in the classroom and for supporting culturally relevant teaching throughout the Pollution Unit. For instance, informal conversations and classroom planning time during the course were valuable in helping Niyah, Hope, and Estelle to plan their lessons, to consider the appropriate content and assessments, and to enhance relevancy for teaching Pollution to urban 4th–5th graders.

The objectives of the Pollution Unit included:

- students will learn about the different components that make up the air, and understand what air quality is;
- students will discover types of emissions that go into the atmosphere as a result of combustible byproducts from manufactured products (i.e., man-made objects including machines/industry, transportation vehicles and home products);
- students will understand the negative effects that these pollutants have on human beings, animals, plants, and environment; and
- students will know how they can help reduce air pollution in their own communities (Estelle & Hope, Unit Lesson Plans).

In Table 2.2 the lesson objectives and the lead teacher responsible for teaching their particular lesson plan for the Pollution Unit are provided (Hope, Microteaching paper).

TABLE 2.2 Summary of Unit Plan Objectives

LESSON #	LESSON OBJECTIVE SUMMARY	"HEAD TEACHER"
1.	Pre-assessment: Students will discuss their ideas about the four learning goals of the Pollution Unit	Niyah
2.	Vaseline Experiment (part 1): Students will prepare an experiment to investigate visible and invisible components in the air	Niyah
3.	Vaseline Experiment (part 2): By collecting, observing, recording, and analyzing data, students will understand visible and invisible components of the air	Hope
4.	Researching Air Pollution: By carrying out group research sessions, students will read, listen to, and analyze a brief written passage to determine the causes and effects of a particular air pollutant assigned by the teacher	Hope
5.	Poster Project (part 1): Students will create posters to demonstrate what they have learned about air pollution and how it effects their community	Estelle
6.	Poster Project (part 2): Students will present posters to demonstrate what they have learned about air pollution and how it effects their community	Estelle

Emergent Assertions of Culturally Relevant Science Teaching

The findings and discussion of the study are written as assertions. The assertions, in italics, make use of CRT principles—quoted from Ladson-Billings (1995a)—within the context of science teacher education. The three large themes or assertions are supported by evidence generated from the data analysis (i.e., microteaching papers, lesson plans, researcher classroom observations, informal conversations, and interviews), and reveal the practices and considerations that the three PSTs used in planning, teaching, and assessing their Pollution Unit and implementing principles of CRT in science. The argument set forth in this study is that in order for students (i.e., students of diverse backgrounds, and the 4th–5th graders at JDES, in this case) to learn in culturally relevant ways, their teachers (i.e., in this case, the three PSTs) must learn and understand the principles of culturally relevant teaching in order to teach in this manner.

Assertion 1. *In order to teach in culturally relevant ways, PSTs must have collaborative support with diverse others in making connections and developing practices to teach science, such that they "experience academic success" (Ladson-Billings, 1995, p. 160), not only for themselves as teachers but also for their students.*

This first assertion pertains to how culturally relevant principles are enhanced in practice when learned with others, or within a community of learners (Wenger, 1998). Niyah, Hope, and Estelle spoke positively about their learning and development as teachers of science to enact culturally relevant teaching practices within their lesson plans. One of the most valuable practices for teaching in culturally relevant ways was group collaboration in which they supported each other in the planning and teaching process. Within the group collaborative process in the methods course, doing observations and their pre-assessments in the school, the PSTs were able to focus on the students and consider their needs during the planning of their unit. As an example, Niyah commented about collaborating and planning the Pollution Unit:

> Collaborating with my group members was also extremely valuable in my planning because we were able to brainstorm as a group what we thought would be best for these students. We had been observing this class [the 4th–5th grade] throughout the semester, so hearing about their observations [Estelle & Hope] was a valuable asset to my planning process. (Niyah, Microteaching paper)

During the planning process in the university classroom, the PSTs had several opportunities to work with each other and the researcher as the instructor of the course. The teachers at JDES were also helpful when the PSTs visited the school. Hope commented on the importance of group collaboration among "diverse individuals" during the planning process:

> Working with [Niyah and Estelle] and bouncing our group ideas with [Professor Mensah], our classmates and the classroom teachers and the science teacher at [JDES] gave me hope about how collaboration with diverse individuals can truly create a diverse curriculum. Having many points of view during the planning process was much better than the idea of having to face it on my own. (Hope, Microteaching paper)

The PSTs felt supported in teaching in culturally relevant ways and making learning about pollution engaging and fun for the students. Niyah felt that collaboration and support were needed in order to teach all students, and that "making science attainable for all students will certainly be a heavy task to accomplish, one that I surely cannot do on my own" (Interview). Niyah also wrote that "one of my needs as a science teacher is the support and guidance from my surrounding colleagues. It is only through this collaboration that a change in science education can be made" (Niyah, Microteaching paper). Both Hope and Niyah understood that collaboration with others was not only important for student learning and their academic success in learning science but also necessary for them to become successful teachers.

Assertion 2. *In order to teach in culturally relevant ways, PSTs must use a language that allows them to elicit student roles that will empower students to want to do and learn science; this includes ways to engage students in the knowledge, language, and skills of science—formally (in school) and informally (at home)—and to make personal connections to science. The goals and content for teaching science must be educationally beneficial, such that PSTs* "develop and/or maintain cultural competence" (Ladson-Billings, 1995a, p. 160) *for the students they teach.*

The second assertion addresses ways that students learn knowledge, language, and skills of science through real-world examples. The examples also connect to the lives of students. Therefore, teachers have to find ways of engaging students in relevant science learning that personally connects to them. For example, as the lead teacher for her 2-day lesson, Niyah conducted the Vaseline petroleum jelly experiment. In this experiment, used to represent "visible and invisible" particles in the air, students put Vaseline on index cards and placed them around the classroom to collect dust particles. While conducting her lesson, Niyah referred to the 4th–5th graders as "scientists." This language or reference to the work of scientists invited students to focus on their experiments and to take on the behavior and practices of scientists. The students became engaged in learning and doing science:

> One thing that I believe helped greatly was calling the students "scientists" and telling them that they will be doing a "special" air pollution experiment. This immediately caught their attention, and they were so excited to earn the role of a real scientist They also displayed a great amount of thought and analysis in deciding where to put their samples of Vaseline, as well as hypothesizing what they believed they will see in the Vaseline the next day. (Niyah, Microteaching paper)

During the observations of the lesson, the researcher took pictures of students making observations of their index cards while conducting the Vaseline experiment. They used tools of science, such as magnifying glasses to analyze the samples they collected. The students were very excited about what they found and completed a data sheet to record their findings and ideas.

Ladson-Billings (1995a) described that teachers who promote cultural competence in the classroom find ways to bridge students' interests to school learning. Niyah, Hope, and Estelle "made a huge effort to figure out ways to make the Pollution Unit culturally relevant for their students" (Hope, Microteaching paper). Making the bridge between science and student interest was a challenge for the PSTs. For example, the researcher and the PSTs discussed during the course ways to make science relevant and connected to the lives of urban youth. We generated many ideas about how to promote science learning and content connections. We also discussed how to broaden our definition of what counts as science curriculum and what it means to teach real world examples in science (Researcher Journal). In teaching the Pollution Unit, these ideas became more concrete.

In lesson four, the PSTs asked students to think about how their Vaseline samples would have been similar or different if they placed them in their bedrooms, kitchens, and bathrooms at home versus the various placements in the classroom. Also during the extension exercise of this same lesson, the PSTs asked the students to think about ways that they, their families, and communities were impacted by the causes and effects of air pollution. As an activity in small groups, the elementary "scientists" were asked to brainstorm ideas on how they can decrease the amount of air pollution at home and in their community. Getting the students to talk about their learning and make connections to science at home, Niyah, Hope, and Estelle promoted science learning in culturally competent ways for their students. Being able to connect school science to self and community required a conscious effort to make scientific knowledge applicable and useful as well as personally relevant Niyah, Hope, and Estelle planned for these ideas and provided multiple ways for 4th–5th graders to develop cultural competence in the context of science. They had to think purposefully about what knowledge, skills, and practices they wanted students to develop, while also taking into consideration diversity issues among their students. Hope wrote:

> [Niyah, Estelle] and I also made a strong effort to incorporate a variety of linguistic, artistic and analytic communication styles into the sequence in an effort to give all children a chance to explore the ideas and concepts of the sequence in many formats. Linguistically we included: verbal discussion (lessons one—five), silent reading (lessons one—four), out loud reading and listening comprehension (lesson four), and written responses—both text and graphic (lessons one—five). We enabled students to communicate ideas through the use of artistic expression including: observational drawing (lesson three), collage and imaginative drawing poster construction (lesson four), creative explanation (story telling) of group posters (lesson five). Finally, we asked students to use a range of methods for analyzing content material including: reading text (lesson four), observing concrete samples (lesson three), reading maps (lessons three and five). (Hope, Microteaching paper)

Finally, Hope stated, "I am so glad that [Niyah, Estelle] and I decided to incorporate such diverse teaching and learning styles" into the planning of the Pollution Unit.

Hope reflected however that she and her co-teachers "could have made the lessons even more culturally relevant" had they taken the class on a trip to the locations included on the community map, or have the whole class take the photographs together as a class assignment where students could draw out sources and effects of air pollution in the community through photography (Microteaching & Interview). These modifications to the Pollution Unit would have increased cultural competence in science, made the lessons more connected to the community and relevant to the lives of the students.

Assertion 3. *In order to teach in culturally relevant ways, PSTs must also include their personal interests and reasons for teaching science content. The goals and content for the lesson must also be culturally and personally relevant and focus on real-world connections, such that PSTs* "develop a critical consciousness through which they challenge the status quo of the current social order" (Ladson-Billings, 1995a, p. 160) *for themselves and their students through science.*

The third assertion promotes CRT that challenges the status quo and places the PSTs in positions to challenge and use their knowledge of science to improve their lives and their communities. Questioning the world around you and learning the content of science becomes personal. Estelle, for her 2-day lesson, focused on "environmental racism" where she wanted "to include a cultural/community perspective" (Microteaching paper) and question the effects of air pollution on students and self. Specifically, Estelle wanted to make connections to pollution within the contexts of East Harlem, where most of the students of JDES lived, and the Bronx, where she lived with her two boys and mother. The idea to focus on "environmental racism" was generated from conversations with the JDES teachers who shared that many of the students in the school have asthma and often miss several days of school due to their asthma. Estelle's two young boys have asthma, and this personal relevance provoked her interest in planning, teaching, and assessing the Pollution Unit from a critical perspective. Estelle supported her reasoning for teaching the topic by stating her purpose and supporting it with research and statistics:

> I want to inform them about environmental racism without actually using those words because one microteaching lesson is not enough to explain the relevance of racism. Therefore my goal was to get students to ask why—Why their community is surrounded by major sources of air pollution, why they or someone they know have asthma, or other related health conditions resulting from air pollution … . According to one study, 'Bronx County has some of the highest rates of asthma in the United States [and] rates of death from asthma in the Bronx are about three times higher than

the national average. Hospitalization rates are about five times higher in neighborhoods in the Bronx. It is estimated that 20% of the children have asthma. Within New York City the disparity in asthma hospitalization rates for asthma in Bronx county and East Harlem are 21 times higher then those of affluent parts of the city. (Estelle, Microteaching paper)

For lesson background, Estelle used the JDES neighborhood and the Bronx to make personally relevant connections to the topic of pollution As a lesson goal, she wanted students to become aware of the issue for their communities and show that the JDES neighborhood and the Bronx have "higher rates of air pollutants, which leads to allergies, which leads to higher asthma cases, which leads to higher asthma attacks" (Estelle, Microteaching paper). Estelle used several resource materials found from Internet searches, websites, and tips to save the environment. We discussed in class how to include information gathered from her research that could be understood by the elementary students. Estelle took digital pictures of the JDES neighborhood and created a poster of air pollution sources as a model for using real-world examples (Researcher Journal). She commented that the pictures "offered students a real connection to their communities" and the pictures were visual representations of community sources of air pollution that had an effect on the students' health (Researcher Journal). During her microteaching, Estelle asked students to use their own zip codes to map how close their homes were to major sources of air pollution (i.e., bus depots, waste treatment plants, waste transfer stations, MTA train yards), and student-groups used classroom computers to research asthma hospitalization rates in New York City. As an assessment of her lesson, Estelle noted that students' written responses "showed how they applied new concepts to their own life experiences" as they questioned the impact of air pollution on their health (Estelle, Microteaching paper).

From the process of planning, teaching, and assessing a culturally relevant science unit, and incorporating principles of CRT, Estelle "became more science literate about air pollution" because she had to research and evaluate her teaching (Researcher Journal). She shared that she wanted to be "confident in her instruction" as the science teacher. During the interview, Estelle talked about how much "time and effort" it took to conduct her research and to plan her lessons. It took her "an hour and twenty minutes" to teach her 2-day lesson as the lead teacher because she had "added the cultural perspective to the unit." The science teacher allowed this time and was patient in letting Estelle complete all that she had planned. The 4th–5th graders were given plenty of time to complete and present their final posters on what they had learned from the Pollution Unit. During the poster presentation in class, students demonstrated understanding of man-made causes of air pollution; negative affects of air pollution on humans, animals, plants, and the environment; and identified several components that make up the air and air quality.

Also assessing students' cultural competence at the end of the Pollution Unit, Niyah stated that as children living and growing up in the city the "real-world examples" that the students were able to make about the topic of pollution was exciting and personally meaningful for them:

> The topic of air pollution and air quality was something that the students could make a connection to. Living in New York City, they were able to give many of the real-world examples around them that showed them that air pollution is prevalent in their surroundings … . I believe that this also made it more meaningful to them because air quality is something that affects us all. I literally saw the excitement in their eyes as they spoke about the types of air pollution they see every day. (Niyah, Microteaching paper)

Additionally, Niyah commented that the Pollution Unit was successful in allowing the students to take action about pollution in their lives: "They also learned, through the following lessons with [Hope and Estelle], that they could take actions to reduce pollution, which made it even more meaningful to them" (Niyah, Microteaching paper). Being able to encourage students to take action is part of challenging the status quo.

For the final microteaching paper, the PSTs had to reflect on their microteaching experience and offer a few examples of student learning. They were asked to construct a science teaching philosophy and write about changes they would make to their teaching and lesson plans based on their microteaching experience and teaching in culturally relevant ways and discuss their professional growth as teachers. The student assessments and philosophy statements, as well as informal conversations and interviews with the three PSTs, were useful in noting the lessons learned and making connections to CRT in pre-service teacher education and assessing their growth as teachers. Some examples were presented above, but more are provided in the following section.

Assessing Science Teaching

Niyah, Hope, and Estelle were reflective of their microteaching experience and what the 4th–5th graders learned from the Pollution Unit. Niyah, from assessing the whole process of planning, teaching, and assessing student learning, commented in the interviews and debriefing sessions several things, such as the role of science and students' misconceptions. She replied that "science teaching should address the needs of diverse learners, and should incorporate the cultures of the students" and that "students need to learn science by making sense out of the world around them. This can best be achieved when they make the connection between science and their everyday lives." Furthermore, Niyah expressed that "implementing culturally relevant teaching was definitely challenging at times because to implement these successfully, teachers need to truly know their students and their backgrounds on an intimate level." Niyah was also honest in sharing her experiences in teaching students of diverse backgrounds. Although she had been around children from diverse backgrounds, being placed at JDES for the semester was her "first time acting as a head teacher to a whole group of diverse students." In learning to be a culturally relevant science teacher, Niyah made connections to her own learning as a student in the methods course:

> Through this experience, I came to gain a deeper understanding of the issues we have been discussing in class throughout the semester. I learned that acknowledging and validating the cultures of our students is crucial in making science meaningful to students. Part of this task is recognizing that as teachers, we are required to make critical decisions in the classroom.

In Hope's assessment of student learning from her lesson plans and microteaching experience, she commented extensively during the interview and debriefing sessions about the success and misunderstandings students had about science. As an example, she developed a graphic organizer for students to complete where they drew pictures, wrote observations, and made analyses of their observations from the science activities as part of her microteaching. She found the graphic organizers to be "a great success" for student learning, and noted that "after seeing students use them first hand, I realize how important visual methods of organization are to promoting diverse student learning." Hope also commented about what student learning looked like across the Pollution Unit. As a co-teacher, she assessed student learning through the poster presentations:

> All students responded to the major concepts of each question on the handout and I think it was because the questions were asked in a clear and visual format. Some students copied the exact phasing from the written air pollution handout, but I noticed that during [Estelle's] lesson when students created posters and presented them that even though they copied phrasing, they still understood the major concepts at the conclusion of the lesson sequence. (Hope, Interview)

Similar to Niyah, Hope learned a great deal from the microteaching experience, particularly about classroom management. Working with the 4th–5th grade elementary students, she commented that good planning became the means for classroom management. From observing her microteaching on the second day of her lesson, Hope was calmer than day 1 where she directed the class in what to do. On day 2 she gave fewer directions and students were able to figure out what to do. She also allowed students to talk and share their ideas more on day 2 than day 1. She worked with the students by allowing them to express their learning through small and large group discussions.

Hope realized "how important diverse teaching and learning styles are to implementing a successful lesson sequence." She commented:

> I learned how much students enjoy becoming scientists and using scientific processing skills and terminology. I watched them as they carried out experiments, observed and recorded data, analyzed and discussed results with each other and finally educated each other about air quality and air pollution. I learned that children are starving for meaningful scientific experiences and I am really glad that in between all the substitutes, classroom changes and testing preparation, [Niyah, Estelle] and I were able to teach them a carefully planned lesson sequence that connected scientific concepts to their everyday lives and promoted conceptual change in their thinking! (Hope, Microteaching)

Estelle also learned much from planning the Pollution Unit. Estelle chose several posters to conduct an evaluation of student learning and her teaching. She brought these posters to the university classroom and shared them with her teacher peers. Estelle commented that "students had developed an understanding of the major sources of air pollution" and that they were able to make connections to "the health effects associated with air pollution within their community" (Researcher Journal). Estelle wrote about one student in particular and assessed his learning:

> [Kevin] drew on his scientific concepts to explain how air pollution affects his community. His presentation and poster illustrated a community perspective. He conceptualized the correlation between air pollution and its affect on living organisms and the environment. He identified one of the sources of carbon monoxide as the [expressway running past his school]. He scientifically conceptualized that heavy traffic contributes to poor air quality and health hazards in his community. Overall, [Kevin] achieved community awareness, he identified a source of pollution, and the impact air pollution has on air quality, and what makes up the air. (Estelle, Microteaching paper)

From coding the microteaching papers of the three PSTs and using other data sources collected for this study, the principles of CRT are evident in the planning process, descriptions of their microteaching experiences, observations of classroom teaching, reflections on student learning and professional growth. In the section that follows is a discussion of the findings of the study. The implications and lesson learned in educating pre-service teachers about culturally relevant teaching in science and the implementation of this approach for teacher education in general are also provided.

DISCUSSION AND IMPLICATIONS

Educators argue for the use of culturally relevant pedagogy to address the learning needs of students in culturally diverse classrooms. Many employ Ladson-Billings' (1995a, 1995b) three principles of culturally relevant teaching, which are: "(a) students must experience academic success; (b) students must develop and/or maintain cultural competence; and (c) students must develop a critical consciousness through which they challenge the status quo of the current social order" (p. 160). I argue that in order for students to learn in the ways that Ladson-Billings outlines, their teachers must learn them similarly so that they may teach in this manner. Therefore, CRT principles are introduced in an elementary science methods course with three pre-service teachers. I discuss Niyah, Hope, and Estelle's experiences in planning, teaching, and assessing a six-day Pollution Unit in a 4th–5th grade urban elementary science classroom in New York City. In this final section, the lessons learned as a teacher educator through the pre-service teachers' learning to implement principles of CRT in science education are discussed. The implications and limitations of the study pose additional ways that CRT can be used in teacher education and the eventual uptake of such practices in diverse, urban classrooms.

Science Teacher Education

Three assertions were introduced in this study that connect theoretically and practically with Ladson-Billings' (1995a, 1995b) principles of culturally relevant teaching for application in science teacher education. First, in order for pre-service teachers to teach in culturally relevant ways, they need support in the form of collaboration with diverse others so that they have success in planning and teaching science in culturally relevant ways. Second, pre-service teachers need opportunities to adopt a language that elicits their roles as science teachers who are empowered to teach in culturally relevant ways, thereby engaging their students in the knowledge, language, and skills of science and empowering their students to be successful learners of science. Third, pre-service teachers are encouraged to find personal relevancy in the science content they are teaching as well as develop a critical awareness of how science can be used to deepen their understanding and application of science to empower and improve their lives and the lives of their students.

Generally, the findings of the study strongly suggest that collaboration and support, personal empowerment to teach, and personal relevance of the subject matter are important practices for teacher education to adopt in the preparation of teachers for diverse classrooms. These ideas connect to previous research such as the six salient characteristics that define culturally responsive teachers (Villegas & Lucas, 2002) and Furman's (2008) demographic tension and effectiveness tension for multicultural teacher education. For science learning in particular, employing the principles of CRT in science teacher education opens opportunities for both teachers and students to move science as an elementary school subject and students traditionally marginalized from learning science to the forefront. This is particularly empowering when science teaching and learning happens within a diverse urban elementary school setting (Mensah, 2011). The three PSTs mentioned how important it is to their success in teaching the Pollution Unit to have the support of diverse others (i.e., their pre-service teacher peers, course instructor, and cooperating teachers at the school) to exchange ideas and to engage in conversations that focus on student diversity, student learning, and pedagogical and content connections. Therefore, safe and supportive learning environments, such as the elementary science methods course and the elementary school placement, are ideal for building pre-service teachers' efficacy to learn and teach in culturally relevant ways.

An additional component for teacher education in implementing CRT is through the microteaching experience—planning, teaching, and assessing a culturally relevant science unit for diverse learners. The microteaching experience increases teacher self-efficacy to teach science (Gunning & Mensah, 2011), and offers a collaborative context to learn and enact the principles of CRT in an authentic classroom setting with students of diverse backgrounds, learning styles, and interests. As teachers learning to teach diverse students, the PSTs interact directly with the students and teachers at JDES. These collaborative interactions are invaluable to the PSTs' growth and development as science teachers. The microteaching experience enables the PSTs to feel success in teaching science so that their students also experience success in learning science. The PSTs and their students learn about pollution in scientifically based ways, increasing their cultural competence about the topic and ability to bridge home and school connections. The PSTs address reasons for learning about pollution and focus on real-world examples so that the topic is engaging and motivating to students, as represented in students' science experiments, classroom discussions, and final poster presentations. The poster presentations underscore the students' academic success, cultural competence, and critical consciousness in learning the goals and objectives of the Pollution Unit. Moreover, the poster presentations solidify Niyah, Hope, and Estelle's success as culturally relevant science teachers and the use of CRT in science teacher education.

Lessons Learned

There are three lessons to share from conducting this study that may be helpful for teacher educators as they also consider curriculum and effective approaches in the preparation of teacher candidates for diverse classrooms. These lessons connect to the broad recommendations for the preparation of teachers for diverse classrooms (see Furman, 2008; Sleeter, 2001; Villegas & Lucas, 2002). First, it is vital that teacher educators develop strong collaborative partnerships between universities and urban schools. Both sites offer theoretical and pedagogical support (i.e., connect theory to practice) for teacher learning, student engagement, and enactment of CRT, particularly in teaching elementary science.

Second, planning, teaching, and assessing curriculum that is grounded in principles of CRT demands a lot of "time and effort" as Estelle comments indicated, about planning and researching to include a cultural component in the curriculum. For PSTs who do not share common cultural knowledge and experiences similar to the students they will teach, additional support, suggestions, and resources will be beneficial in helping them to make relevant connections. As previously mentioned, spending time in diverse classroom settings, making observations, working with small groups of students, and discussing ways to meet the needs of diverse students, are suggested as essential elements in a methods course. Moreover, pre-service teachers need ample time to plan, research, discuss, and teach culturally relevant curriculum throughout their teacher education preparation. This is especially required for topics that raise social consciousness and topics that are not traditionally addressed in the school curriculum. Niyah, Hope, and Estelle—as student observers who had not yet done student teaching—only touch the surface in talking about environmental racism in their Pollution Unit. However, within the one methods course, an early foundation to think about science curriculum through a social action/social justice or critical lens is introduced to Niyah, Hope, and Estelle. They are able to implement some initial understandings of what critical consciousness looks like and the possibilities it holds to engage students in science.

Finally, the last lesson addresses science as a priority subject in the elementary school program. Science in many elementary school settings is not a priority subject (Spillane et al., 2001; Tate, 2001). In many cases, the elementary school program is overtaken by disempowering policies and institutional barriers that inhibit real engagement in science teaching and learning (Carlone, Haun-Frank, & Kimmel, 2010; Mensah, 2010; Rodriguez, 2010). For PSTs to teach in culturally relevant ways in science classrooms, and for their students to learn science in a similar manner, then science in the elementary school must be given the time, resources, priority, and attention it deserves so that students experience academic success, develop and maintain cultural competence, and develop a critical consciousness through which they challenge the status quo of the current social order. Teaching science in the elementary school is a challenge to the status quo.

LIMITATIONS AND CONCLUSION

Although this study takes place in one section of a graduate level, elementary science methods course, with a small group of diverse PSTs, the findings suggest that fostering collaborations and partnerships, structuring classroom time to implement CRT in classrooms and in teacher education are valuable approaches in the preparation of all teachers for diverse classroom settings. In addition, the ideas explored in this article indicate that classrooms where this type of teaching is taking place requires that teachers experience learning in ways other than we expect them to teach. But, how will our teacher candidates teach in the ways we advocate if they have not experienced teaching and learning in these ways themselves? Villegas and Lucas (2002) commented that as teacher educators we should expect our PSTs "to demonstrate an initial ability to tailor their teaching to particular students within particular contexts, a central quality of culturally responsive teaching" (p. 30). Accordingly, pre-service teachers need sufficient opportunities to think, plan, teach, and assess their growth and development as culturally relevant teachers. In this study, science teacher education is suggested as one context. They realize that teaching for academic success, cultural competence, and critical consciousness is complex and requires collaboration and practice. As teacher educators, we serve as supporters and advocates for our pre-service teachers' on-going professional development. In addition, we need thoughtful, well-planned teacher education curricula and activities that educate our pre-service teachers in learning what CRT is, and its potential usefulness not only in their classrooms but also for their education. Therefore, this study extends to readers the potential of culturally relevant science teaching within teacher education and its application in elementary science classrooms.

As a teacher educator, my hope is that my pre-service teachers carry models of teaching, foundational practices, and educative experiences from teacher education with them into their teaching careers, and build on their initial understandings of what it means to teach students of diverse backgrounds. However, they must experience this in positive, collaborative, and supportive teacher education programs, through courses and school–university partnerships that prepare them for increasingly diverse classroom settings. The contribution that this study offers is how and what it may look like in one elementary science teacher education methods course that is focused on the preparation of elementary science teachers for diverse, urban classrooms.

REFERENCES

Carlone, H. B., Haun-Frank, J., & Kimmel, S. C. (2010). Tempered radicals: Elementary teachers' narratives of teaching science within and against prevailing meanings of schooling. *Cultural Studies of Science Education, 5*, 941–965.

Charmaz, K. (2006). *Constructing grounded theory: A practical guide through qualitative analysis.* Thousand Oaks, CA: Sage.

Creswell, J. W. (2007). *Qualitative inquiry and research design: Choosing among five approaches* (2nd ed.). Thousand Oaks, CA: Sage.

Department of Education. (2010). *School profile data.* Retrieved from http://schools.nyc.gov/Accountability/tools/report/default.htm

Furman, J. S. (2008). Tensions in multicultural teacher education research: Demographics and the need to demonstrate effectiveness. *Education and Urban Society, 41*, 55–79.

Gay, G. (2002). Preparing for culturally responsive teaching. *Journal of Teacher Education, 53*, 106–116.

Guba, E. G., & Lincoln, Y. S. (1989). *Fourth generation evaluation.* Newbury Park, CA: Sage.

Gunning, A. M., & Mensah, F. M. (2010). One pre-service elementary teacher's development of self-efficacy and confidence to teach science: A case study. *Journal of Science Teacher Education, 22,* 171–185.

Ladson-Billings, G. (1994). *The dreamkeepers: Successful teachers of African American children.* San Francisco: Jossey-Bass.

Ladson-Billings, G. (1995a). But that's just good teaching! The case for culturally relevant pedagogy. *Theory into practice, 34,* 159–165.

Ladson-Billings, G. (1995b). Toward a theory of culturally relevant pedagogy. *American Educational Research Journal, 32,* 465–491.

Mensah, F. M. (2010). Toward the mark of empowering policies in elementary school science programs and teacher professional development. *Cultural Studies of Science Education, 5,* 977–983.

Nieto, S. (2004). Affirming diversity: The sociopolitical context of multicultural education (4th ed.). Boston, MA: Allyn and Bacon.

Rodriguez, A. J. (2010). Exposing the impact of opp(reg)ressive policies on teacher development and on student learning. *Cultural Studies of Science Education, 5,* 923–940.

Sleeter, C. (2001). Preparing teachers for culturally diverse schools: Research and the overwhelming presence of Whiteness. *Journal of Teacher Education, 52,* 94–106.

Spillane, J. P., Diamond, J. B., Walker, L. J., Halverson, R., & Jita, L. (2001). Urban school leadership for elementary science instruction: Identifying and activating resources in an undervalued school subject. *Journal of Research in Science Teaching 38,* 918–940.

Strauss, A., & Corbin, J. (1998). *Basics of qualitative research: Techniques and procedures for developing grounded theory.* Thousand Oaks, CA: Sage.

Tate, W. (2001). Science education as a civil right: Urban schools and opportunity-to-learn considerations. *Journal of Research in Science Teaching, 38,* 1015–1028.

Villegas, A. M. & Lucas, T. (2002). Preparing culturally responsive teachers: Rethinking the curriculum. *Journal of Teacher Education, 53,* 20–32.

Wenger, E. (1998). Communities of practice: Learning, meaning, and identity. New York: Cambridge University Press.

SECTION I   QUESTIONS TO PONDER ...

1. What is constructivism?
2. Why is constructivism important in learning science?
3. Why is it important to make science education culturally relevant?
4. What is a rationale for teaching science in a multicultural way?
5. Why must one be a scientifically literate person?

Section II

BASIC TOOLS OF SCIENCE EDUCATION

Objectives

- Determine the uses for the science process skills.
- Explain the difference between an observation and an inference.
- Differentiate between an activity and an experiment.
- Analyze the 5E lesson plan model.

Key Concepts

science process skills: The tool or skill used to do science.

experiment: A controlled comparison where an independent variable is manipulated to test a hypothesis.

5E model: Constructivist lesson plan model based on the learning cycle. The phases of the model are engagement, exploration, explanation, elaboration, and evaluation.

Introduction

This section of the book introduces some of the basic tools for engaging in and teaching science. The process skills are the tools for how all science is performed. Learning the process skills and being able to manipulate them is essential to the scientific process. The process skills are a part of all the science activities completed in the classroom. Once a student has a handle on the basic process skills, learning the intergrated process skills should be simple. Subsequently, this knowledge can be utilized while performing experiments.

It is important to keep in mind that not everything that happens in a science class is an experiment. Most are science activities and not experiments. Experiments have a specific definition in science. If you are testing a hypothesis in a controlled manner by manipulating variables, you are performing an experiment. If you are not testing a hypothesis and not doing multiple trials, you are most likely doing a science activity. All experiments are activities, but not all activities are experiments.

Science teachers have many lesson plan models to choose from; however, the 5E model is a constructivist model developed specifically for science. It is integrated and is great for completing thematic units.

READING 3

BASIC SCIENCE PROCESS SKILLS

OBSERVE, INFER, AND CLASSIFY

By John Settlage and Sherry Southerland

HIGHLIGHTS

- Process skills represent the active doing of science and provide benefits to the classroom that extend beyond science learning. Having all students observing, inferring, and classifying allows them to engage in the actions of the science culture.

- Observing is the most fundamental of the basic science process skills and includes sight as well as the other senses. Students are to learn that observations should be unbiased and valued for their factual basis.

- Inferring is an attempt to explain the reason or cause for what has been observed. Helping students appreciate that inferences are judged by how sensibly they explain observations will allow them to distinguish between effective inferences from less suitable efforts.

- When classifying, scientists organize objects into distinct categories. The keys to an effective classification system include using observable properties and always separating objects into only two categories at any one time.

- The culture of science involves very special actions that have been formalized as it has been practiced. Recognize that teaching students includes helping them appreciate that scientific thinking is not automatic or natural. Instead, classifying and other scientific actions represent one group's way of viewing the world.

- Teachers should reserve teaching about the integrated science process skills for students who are in upper elementary school or middle school. The basic science process skills provide a foundation for students and teachers who are ready to tackle integrated process skills such as manipulating variables and posing operational definitions.

TEACHING ALL FEATURES OF SCIENCE

Too many people regard science as a collection of facts and formulas. As important as scientific explanations are to science, studying science requires more than just learning about the products of science. Teachers who are unaware of the actions of science hold an incomplete view about the science learning their students should experience. Presenting science to students as an accumulated body of knowledge addresses only one aspect of the broad culture of science and will give students an inaccurate and limited view of the discipline. Teaching science with too much emphasis on the content would be like teaching language arts by providing students with a few nouns but no verbs—and expecting them to construct sentences. Likewise effective science teaching incorporates a healthy balance of concepts and skills.

Most scientists are not attracted to their chosen profession because they want to memorize the theories and discoveries of other scientists. Instead, people are drawn to science because of the things they get to DO: the actions of science that lead to scientific discoveries. The appeal for those who become science enthusiasts is the opportunity to inquire about their world. Emphasizing inquiry as an essential component of the actions of science is particularly important when teaching in diverse settings. By allowing students to inquire about the world in or around the classroom, teachers accomplish several important educational goals.

Teaching students to inquire is fundamental to helping them become participants within the culture of science. However, the unique features of scientific inquiry are not immediately apparent to all students. Furthermore, not all students enter the classroom with the same cognitive, behavioral, and physical abilities to participate in inquiry. Although inquiry is a centrally important feature of the actions of science, careful support is required to help students become successful within these actions. We will focus on a set of specific actions, called the **science process skills**, which lay the foundation for scientific inquiry.

BENEFITS OF PROCESS SKILLS

Achieving an appropriate balance is one of the secrets of becoming a professional. For example while a group of American Indian college students were preparing to become social workers, nurses, and psychologists, they were challenged to balance their cultural perspectives with Western, scientific perspectives (Weaver, 2000). Being successful in courses meant that they needed to learn a different way of thinking. But to be effective when working with American Indian clients, they needed to balance this new thinking with their native culture.

Duffy (1998) compared teaching to maintaining one round stone on top of the other—a balancing act that requires constant attention and frequent adjustments. Teachers, in terms of supporting their students' science learning, are challenged to achieve a balance between science concepts and process skills. Too much content can stifle student interest. Devoting too much attention to the process skills can distract students from learning the substantive ideas within science. You might imagine the pull in opposite directions: to one side is the attraction of having students actively involved in working with materials while to the other side is the desire for students to master essential scientific concepts. As classroom teachers, we have to find a way to avoid pulling too far in one direction or the other.

There are several reasons why science process skills should be present within elementary and middle school science lessons. Teachers may feel pressured to not spend too much energy on having students use the science process skills. For example students are more active and talkative when they are using the process skills than when they are reading from a textbook. Sometimes people confuse active learning with messing about and there are instances where this may be the case. But when the use of science process skills has a clear purpose, to the teacher and to the students, then they aren't a matter of playtime. What follows are several justifications that support having students engage in the science process skills as a regular component of science instruction and as a necessary feature of science learning.

Sense-Making Tools

One key to closing the academic achievement gap is to make certain that students are learning both the content and the processes of subjects (North Central Regional Educational Laboratory, 2004). Frequent and increasingly challenging uses of the science process skills support students in developing their efforts at scientific inquiry. This is important because increased intellectual development diminishes students' dependence on the teacher in the classroom. As the students become more skilled and confident with doing science, they begin to make connections for themselves. The teacher is still an important resource to the students but is no longer the source of all knowledge. As a consequence the students develop the power to become more independent learners. Simply learning the process skills does not magically cause this to happen. Instead students need to actively engage science process skills as tools to effectively expand their understanding of science. With each use of the tools, and with increasingly sophisticated application of the tools, students are better able to make sense of the natural world.

Supporting Language Development

Science process skills also support the development of student language because, as a part of using these tools, students are simultaneously called on to engage in discussions with others. The need to communicate what is being seen or to describe ideas to another person challenges a child to articulate his or her thoughts. For English language learners, the opportunity to practice oral communication in the context of actual science activities is a powerful way to develop fluency (Amaral, Garrison, & Klentschy, 2002; Linik, 2004). The same support of language fluency applies to young learners. The need to provide descriptive words and complex ideas creates the need to find specific and elaborate ways of using language. Students can practice the same skills when they communicate to each other about what they have read, but there is a heightened sense of purpose when students are more actively engaging process skills and communicating ideas about real materials. As a result students develop greater control over their abilities to communicate and are less dependent on the teacher for their learning. More detailed information about strategies for supporting process skills development is in the chapter by Maatta, Dobb, and Ostlund (2006) in the book *Science for English Language Learners* by Ann Fatham and David Crowther.

Creating a Community of Learners

Science process skills also provide the opportunity to create a community of learning within the classroom. Students are engaging science process skills when they are working with materials, ideas, and other people. In other words, this practice replicates the social aspect of a professional scientific community: information is exchanged, explanations are ventured, and understandings are negotiated. In Chapter 1, we identified three myths about teaching, and here we see a way to undercut and replace all three. Because instilling a collective effort to understand science makes the class less dependent on the teacher, the teacher no longer needs to meet the demands of serving as the full-time expert (Dawes, 2004). Through shared experiences, individuals from an array of backgrounds can jointly participate in lessons rich with the science process skills.

Fostering Natural Curiosity

A final justification for relying on the science process skills is the potential to make use of students' natural curiosity. When people are intrigued by equipment or an artifact from the natural world, the challenge of motivating them is almost solved. It's hard to envision elementary or middle school students asking, "Why do we have to learn this?" when they are engaged with materials and using science process skills. It is true that the materials can be a source of distraction when the teacher tries to shift the students' attention to important information. But this is a much more solvable problem than trying to engage students in the first place. Although words such as *enchantment* and *wonder* seem almost too dreamy to apply to science teaching, the opportunity to engage with science through direct experiences does seem to have a somewhat magical quality. Not all children will be motivated to the same extent for every activity. Yet the motivation to learn is more likely to come from inside the students when they are using process skills than when they need to rely on someone else to communicate information to them.

OBSERVING

Several years ago a few educators tried to emphasize the active dimension of science by turning it into a verb. Even though the word *sciencing* hasn't persisted, the premise was a good one. If we want science to become more accessible to a wider spectrum of students, we need to make it clearer that science study requires doing science, not simply learning about science.

We suggest using the verb form for each of the basic science process skills. We prefer **observing** to *observations*. Observations have already been done; observing is something that we can and will do ourselves. Commonly, observing means using our eyes to make observations, and that is true much of the time. However, scientists make observations with their other senses, and students should be guided to recognize when they are using their senses while observing. For students with particular physical limitations, the teacher will need to make appropriate accommodations. This includes allowing the student with poor vision to listen to the sounds of different types of powders as they are poured into a container or to use his or her sense of touch to observe the texture of rock samples. In a similar vein, the varied ways of observing (seeing, tasting, touching, listening, and smelling) are important to employ with students with limited cognitive capacities. Relying on multiple modes of observing allows individuals to build a more complete understanding of their experiences. Furthermore, encouraging students to use a variety of senses while observing benefits English language learners. With more opportunities to use the language, they have increased opportunities to develop and expand their vocabulary.

As we advance the notion that observations can make use of all of our senses, we have the major challenge of avoiding the urge to interpret the significance of observations too quickly. Observing should focus on telling "what it is" and "how it is" but not "why it is." We have found it convenient to regard observations as facts. This means that observing shouldn't differ depending on who makes the observation. Observing in science is an active endeavor and should be done with care. Two people, one carelessly looking about and one carefully observing her surroundings, may be seeing the same thing but not noticing the same features. Sherlock Holmes demonstrates in the following dialogue the distinction between seeing and observing. When one person makes a careful observation, such as the number of steps in a stairwell, and another person can confirm that she can witness the same thing, then we are talking about the process skill of observing.

> Sherlock Holmes and Watson discuss the difference between seeing and observing (from "A Scandal in Bohemia" by Arthur Conan Doyle [2004]):
>
> *Watson:* And yet I believe that my eyes are as good as yours.
> *Holmes:* Quite so. You see, but you do not observe. The distinction is clear. For example, you have frequently seen the steps which lead up from the hall to this room.
> *Watson:* Frequently.
> *Holmes:* How often?
> *Watson:* Well, some hundred times.
> *Holmes:* Then how many are there?
> *Watson:* How many! I don't know.
> *Holmes:* Quite so! You have not observed. And yet you have seen. That is just my point. Now, I know that there are seventeen steps, because I have both seen and observed.

Differentiating Fact from Opinion

Before proceeding we need to clarify the role of opinion within science. As will become apparent, such traits as creativity and intuition are very important to science. We don't want to dismiss the role of individual insights, because they are absolutely necessary for transforming information into understandings. Nevertheless, we need to help students regulate the use of their opinions while they are doing science. Facts and opinions are distinct yet interdependent ideas. To a large extent scientific inquiry relies on facts, gathered through observing, as its starting point. When we interpret facts too quickly, we may neglect other observations that could be significant—and, in turn, prematurely come to a conclusion that is incorrect. By emphasizing the idea that observing leads to facts, we want students to avoid the danger of allowing preconceived ideas from influencing what they notice.

In reality it's impossible to expect scientists, either as professionals or as students, to do all their observing first and then make the switch to explain these facts. But a distinction needs to be made between observing and explaining. To aid students in this distinction, ask, "Which of your senses did you use to make that observation?" If they cannot identify a sense, then the statement is likely an inference.

Observing as Paying Attention

There is a legend about zoologist Louis Agassiz and his strategy for teaching his students about observing. Agassiz founded Harvard's Museum of Comparative Zoology and apparently had the habit of putting his new students in front of a preserved fish and telling them to observe (Menand, 2001). Then he'd leave them with neither tools nor hints about how to proceed. Nathaniel Shaler, who would eventually become a professor of paleontology at Harvard, recounted one version of Agassiz's teaching technique:

> When I sat down before my tin pan, Agassiz brought me a small fish, placing it before me with the rather stern requirement that I should study it, but should on no account talk to anyone concerning it, nor read anything relating to fish until I had his permission to do so. To my inquiry, "What shall I do?" he said in effect: "Find out what you can without damaging the specimen: when I think that you have done the work, I will question you." (Shaler, 1946, p. 213)

Ultimately, after Shaler spent many hours observing and documenting, Agassiz was sufficiently pleased by his efforts. Another of Agassiz's students, who eventually distinguished himself with a career as an insect expert, related his story about observing a fish as part of his training:

> In ten minutes I had seen all that could be seen in that fish, and started in search of the professor, who had, however, left the museum … . Half an hour

passed, an hour, another hour; the fish began to look loathsome. I turned it over and around; looked it in the face—ghastly; from behind, beneath, above, sideways, at a three-quarters view—just as ghastly. I was in despair; at an early hour, I concluded that lunch was necessary; so with infinite relief, the fish was carefully replaced in the jar, and for an hour I was free.

On my return, I learned that Professor Agassiz had been at the museum, but had gone and would not return for several hours. Slowly I drew forth that hideous fish, and with a feeling of desperation again looked at it. I might not use a magnifying glass; instruments of all kinds were interdicted. My two hands, my two eyes, and the fish; it seemed a most limited field. I pushed my fingers down its throat to see how sharp its teeth were. I began to count the scales in the different rows until I was convinced that that was nonsense. At last a happy thought struck me—I would draw the fish; and now with surprise I began to discover new features in the creature. Just then the professor returned.

"That is right," said he, "a pencil is one of the best eyes." (Scudder, 1879, p. 450)

We don't want to give the impression that good teachers should make students learn how to observe by forcing them to stare at a dead fish for, at least with some of Agassiz's students, days at a time. The reason we present these stories is to illustrate that observing is much more involved than many people realize. One needs to observe with great care and attention to details, ideally writing or drawing what is witnessed. These two future scientists both remarked that as a result of this experience, they came to look at something ordinary with a fresh perspective.

Observing with Minimal Bias

Observations are supposed to be free from bias. This means that *what* one expects to observe shouldn't have much influence. By taking steps to avoid allowing our own opinions to obscure what we observe, we are approaching **objectivity.** What can be done to help improve the quality of observations? One way is to have multiple people make observations and compare what is found. If we know our observations will be checked against others' observations, we will probably be more careful because we'd rather not find that the statements once thought to be observations are in reality just opinions.

In other words, it is likely that observations will be more reliable if others check them and that they will be less influenced by personal prejudice if a large number of people are recording them. Ideally we would be able to re-perform certain experiments with more people making observations and then check to see whether the results are the same. This desire to eliminate prejudice in observations ties into the scientific habit of mind of skepticism. Think about watching a magician as she performs a sleight of hand trick. When we can't believe our eyes, one of our inclinations is to want to see the trick again. It is as if we doubt that what we witnessed really happened. The same idea applies with regard to repeating observations.

Observing within Learning

We have given the impression that students' use of the science process skills should begin with observing. However, observing is not simply the first step, and its use does not end after observations have been made. Observing is like breathing: you cannot do it only once. When put to effective use, the process skill of observing continues, just as with breathing, throughout all subsequent activity. This should reinforce the value in having students learn to make unbiased, multisensory observations: they will continue to rely on this skill during all phases of the science lessons and activities.

> Observation exists at the beginning and again at the end of the process: at the beginning, to determine more definitely and precisely the nature of the difficulty to be dealt with; at the end, to test the value of some hypothetically entertained conclusion. (Dewey, 1910/1991, p. 77)

As John Dewey explained, observing is put to use at different times within scientific activity. Students will observe specimens and materials when they are first introduced to them, they will continue to observe as they note patterns, they will make observations as they modify the conditions, and they will observe yet again to determine whether their expectations have been met. None of the process skills can be treated as items on a checklist that can be marked off once they've been used. Younger students can refine their observing by relying on multiple senses and attending to distinctive features, while older students can begin to employ tools to aid in their observing while seeking patterns and regularities as well as unique characteristics (Harlen, 2000).

INFERRING

If observing is the beginning of scientific investigation, then inferring follows very close after. It is a challenge for students to keep facts separated from opinions because students can be very quick to explain what happens. Without having made thorough observations, by being in too big a hurry to explain, students can overlook details that might be crucial. Think about observing as the first stepping-stone in your path of scientific inquiry. Inferring represents the second step. When anyone neglects observing, they bypass the first stepping-stone. Quite literally this describes "jumping to conclusions." One of the most powerful aspects of helping students learn the actions of science is in teaching them how to not jump to conclusions. Careful, detailed, and even written or illustrated observations are key prerequisites to making inferences.

The philosopher John Stuart Mill claimed that inferring is "the great business of life" and each of us is constantly trying to make sense of the world. It's rare for humans to be willing to let the facts stand on their own. Especially when we observe something that is unusual, we almost can't help ourselves from trying to explain what we have witnessed. Rushing to conclusions can cause us to miss key features and, consequently, make inferences that are insufficiently supported by facts. Or one might make an inference that proves to be a dead end—and then need to make more observations to reach a new conclusion anyway. Starting with extensive observations increases the chance the resulting inferences will be justified and accurate.

> To draw inferences has been said to be the great business of life. Every one has the daily, hourly, and momentary need of ascertaining facts which he has not directly observed; not from any general purpose of adding to his stock of knowledge, but because the facts themselves are of importance to his interests or to his occupations. (Mill, 1884, p. 5)

Our definition of **inferring** is: developing an explanation that is based on and supported by valid observations. Synonyms for *inference* include *speculation, suggestion, supposition,* and *hypothesis*. Unlike observations, inferences are drawn from experiences and ingenuity. Someone who is skilled at inferring is able to take observations and then generate reasonable explanations for them.

The Criteria for Judging an Inference

Observations, which are based on use of the senses, can be evaluated by considering whether they represent facts. The test of an inference is much less obvious than the criteria for judging an observation. An inference is an attempt to explain, it is a claim about why something is the way that it is. Inferences are not facts but opinions informed by facts. Unlike opinions, which we can judge based on whether we agree with them, judging an inference forces us to rely upon the supporting facts. Deciding whether an inference is appropriate requires a judgment call, and learning to make useful inferences is again an essential action of the culture of science. We must learn to judge the quality of an inference based on how well it is supported by the observations.

You might use a courtroom as your mental model. The prosecution and the defense express their particular inferences. Both try to convince the judge or jury that their inference makes sense, and they use evidence to support their claim. Pieces of evidence, in the form of photographs, weapons, or tape recordings, are equivalent to observations. Everyone has access to those facts and pieces of data. The job of the lawyers is to use those pieces of evidence to create explanations that are compelling and convincing. The better inference is the one that makes the best sense in explaining the greatest number of observations.

There are many ways in which this analogy between the legal process and science is insufficient. First of all, in science there aren't necessarily two inferences in opposition to each other. In science we may have multiple inferences or hypotheses that need to be considered. Within science there isn't a single judge to decide what inference is correct. Instead science consists of a large, seemingly quarrelsome community that determines the correctness of knowledge claims. The best inference is the one accounting for more of the observations than any other.

In science there isn't only one occasion to make a decision. In our legal system a person cannot be forced to stand trial a second time if someone doesn't like the results of the first trial. Unless things are complicated by appeals, the outcome of a trial is the final decision. In contrast (remember openness to new ideas as a scientific habit of mind?) scientific inferences are ALWAYS open to reevaluation by the broader community of science. When more evidence has been gathered, it can be used to test whether the current inference is still sufficient. As an alternative, a scientist can propose a new inference, and it always remains as a possibility that the old inference can be discarded because the new one does a better job of explaining the data.

Another problem with the courtroom analogy is the inevitable appeal to emotions. In a courtroom, attorneys will often attempt to sway opinions by trying to take advantage of people's sympathies, by taking pity on either the victims or the accused. In judging inferences we must teach students to rely on the quality of the observations for their decision. In addition, students must learn to resist being influenced by such things as the personality of the individual who champions an inference.

One way to avoid allowing biases to interfere with the consideration of an inference is to try to detach the claim from the person making it. An inference is an idea that is held out for everyone to consider. It should withstand a comparison with the available observations. When an inference is judged to be unreasonable or unacceptable, it shouldn't be seen as a reflection of the person who proposed it. An inference doesn't have feelings, and so there should be no worries when one is rejected. This is particularly important to model for your students.

A teacher should make it clear to his or her students that all inferences need to be carefully examined, even (or especially) ones proposed by the teacher. By doing this the students will learn to understand that part of the actions of doing science is to critically examine inferences. Polite critique of an inference is not a sign of disrespect but a central and defining feature of the scientific culture. As such, questions such as "How do you know that?" or "What is your evidence?" should be commonplace in a classroom. Furthermore, these questions ought to be posed by teachers and students. Such questions will feel more natural for some students than for others, because many cultures are taught that questioning an adult is inappropriate. Because asking "how we know" in science is such an important scientific action, explicit conversations and repeated modeling are necessary to help students recognize the value. In addition, the teacher must provide support and enthusiasm for students when they pose such questions. Over time, the need to support all knowledge claims with convincing evidence can become a natural part of a vibrant classroom culture.

So What's the Right Answer?

Because scientific explanations always have the potential for being replaced by newer, better explanations, it is never accurate to say that a particular inference has been proved. Even though many people anticipate that science will provide the right answers and that the answers will hold true for all time, in actuality scientific explanations may be replaced. There are a host of examples of this happening throughout the history of science. For example biologists once thought that the number of predators, such as foxes, controlled the number of prey, such as rabbits, in an area. This made sense because the changes in rabbit and fox populations seemed to be related. But after a great deal of study of these animal populations, scientists found that rabbit populations were influenced by the amount of available food and that this in turn influenced the number of predators that could survive in the area. This is an example of an inference re-formed because of a fresh interpretation of observations.

FOR REFLECTION AND DISCUSSION

The scene illustrated in Figure 3.1 is an open field of snow. The dark marks seem to be tracks made in the snow. Generate a list of statements about this scene. Some of these statements should be observations and some should be inferences. Write these as statements, but without including a form of the word observe or infer in those sentences. Ask someone to identify which they regard as observations and which they suspect are inferences. What were the criteria they used to help them to decide?

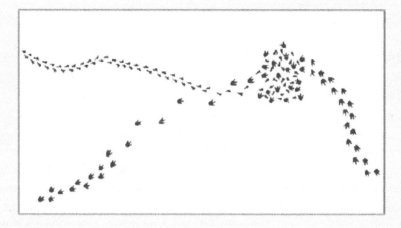

FIGURE 3.1: This drawing of animal tracks allows you to practice observing and inferring.

This changing of explanations is another example of the culture of science. Scientific explanations are always reasonable enough for the length of time that the community accepts them. But it would be a strange strategy to try to trap a scientist into saying whether he or she is 100 percent confident in an explanation, because the concept of certainty in scientific knowledge is foreign to the culture of science. The current debate about the "greenhouse effect" is a recent example. There may never be sufficient data to claim, with absolute certainty, that the burning of fossil fuels is changing the earth's temperature and climate. However, gathering and analyzing data can lead scientists to make stronger inferences that, in turn, could give greater credence to either side of the argument. The professional scientific community evaluates inferences based upon the strength of the supporting data. So too will students in the classroom be expected to supply observations in support of the inferences they propose.

Science as a Creative Endeavor

In his book *How We Think,* John Dewey (1910/1991) described how inferring connects to observing. He wrote that thinking requires us to take the facts that have been gathered and then make a leap to try to explain the facts. The facts, our observations, are what we use in science as the bases for building observations. A person who is creative is more likely to propose many inferences compared to someone who takes things at face value. Scientific study benefits by having many inferences to consider, because multiple inferences provide more options that may lead to a very powerful explanation. Being creative doesn't mean that any idea is acceptable. There are certain guidelines that must be followed, one of the most important being that one must rely on observations to support inferences. Yes, it is useful to remain open to all ideas, and we shouldn't be too eager to discard inferences prematurely. But after the wild speculation has run its course, we must sort through the assorted inferences to determine how well each is supported by our observations.

Another key to proposing inferences is to avoid allowing one's biases to interfere. John Dewey identified the move from what is known (observations) to what is not yet known (inferences) as being "peculiarly exposed to error." He went on to identify possible causes for these errors: previous experience, self-promotion, strong opinions, false expectations, and good old mental laziness. If those doing science, and this includes professional scientists and elementary school and middle school students, remain attentive to these dangers, then they can use their creativity to help sort through what might otherwise be a confusing jumble of observations. Ideally a teacher would elicit a wide variety of explanations and then guide students to evaluate the viability of these inferences based on the available data. Those inferences that will become the most compelling will account for the greatest number of observations. Although that may not lead to the "right" answer, the process of sorting weaker inferences from stronger ones is important to demonstrate and make obvious to students.

CLASSIFYING

We have established that the culture of science anticipates that there are patterns existing in nature. The goal of scientists is to uncover those patterns, beginning with organizing observations. Unless observations are organized, the task of identifying patterns is difficult. **Classifying** is the process of organizing objects into groups based on observable properties. Through classification, patterns that may not have been apparent when objects were viewed as a large group can become evident.

Before going further, we want to reinforce the idea that classifying and all the other science process skills are activities in which we want students to engage. If you hear the word *classifying* and have visions of how living things are organized (e.g., kingdom, phylum, genus, species) then you know only part of the story. Although existing classification systems are important to scientific literacy, it is also important for students to learn *how* to construct classification systems. In elementary and middle schools, it is valuable for students to become acquainted with classification systems for clouds, landforms, and vertebrates. Beyond helping students make conclusions, the process of classifying allows them to appreciate how current systems for classification came into existence.

Part of the problem with the way that classification is commonly presented in classrooms is that it is too often treated as an end product rather than as something that can provide insight into the actions that led to its production. Students should be taught that any current classification system is only the most recent in a prolonged process of pattern seeking and sense making. In other words, knowing *how* to classify and having ample opportunities to develop classification skills are as important as knowing *about* classification.

Classification in Daily Life

All students should be given the opportunity to build their science knowledge on their personal experiences. The differences among students in a classroom, whether cognitive, cultural, or linguistic, need to be appreciated by the teacher. This means more than simply recognizing that all children are special in their own ways. For every student to become skilled in science, students need to connect their everyday lives with the study of science. This is not to suggest that some children have weaker backgrounds than others, as the reality is that every child comes to school with a rich set of life experiences. Good teachers will recognize those experiences as important foundations for science learning. Too often a teacher will assume that his or her experiences parallel those of his or her students. For example a colleague was teaching a science lesson in an inner-city school. In an attempt to describe kinetic and potential energy, he tried using the analogy of a waterfall. Having grown up in the country and traveled extensively, he had a strong image of waterfalls. When his explanation failed to advance his students' understandings, he realized that his students did not share that experience, and he had failed in his choice of examples. For teachers who embrace the value of student diversity, they will need to work a little harder to uncover what represents "everyday" for students. In this way, science instruction starts with objects and actions with which the students are familiar and comfortable. Starting with the familiar makes it much more likely the students will become comfortable with the objects, actions, and ideas of science.

Life is very complicated, so we try to organize it. We organize our clothes by some system, maybe sorting the dressy clothes from those that are more casual. Likewise we rely on classification systems that others have created to make our lives easier. For example stores are organized so that certain products are all in one section. These are examples of classifying. Essentially there are a bunch of objects that are placed into groups based on selected properties.

The items in a grocery store are classified, in part, by the type of container: canned goods in one section, fresh fruits in a different section, and so on. There are many ways to classify the products in a grocery store. Going to a new grocery store is often a challenge, because you don't yet know their particular classification system (where do they keep the cornstarch?). Once you understand the rules that have been used to classify, then you have no problem locating the products you need. To a large extent, the science process skills are based on very natural human behaviors. We observe our surroundings and try to make sense of what we notice by generating inferences. When we are faced with an assortment of objects, we may try to manage the confusion by creating a system for organizing: junk mail versus personal mail, junk food versus healthy food, reality shows versus documentaries—all of this involves classifying. In this way, learning to classify builds on thought processes students may bring from home. We can expect that studying science will refine and expand students' general classification abilities.

The process of classifying allows for considerable personal choice, yet there are certain standards that make some systems better than others. In what follows, you will learn guidelines for creating effective systems of classification. You might think of these guidelines as a system for separating the good from the weak when it comes to classifying. Nevertheless you can exercise a great deal of flexibility, creating several classification systems for a set of materials. If the guidelines are adhered to, each classification system should be equally effective.

Relying on Observable Properties

When classifying, you need to organize the objects or events based on properties that can be observed. We will refer to this as Guideline #1. Color, shape, weight, and composition are observable properties and are acceptable classification criteria. Inferences do not have a place in classifying. In the scientific culture, we don't classify according to whether objects are funny, scary, or boring. We may classify movies this way, but in science we focus our classifying on the materials as they are, not what we feel about them. Likewise you should not classify according to the usefulness of an object. The process skill of classifying reinforces the need to make observations that are factual, and the best properties to be used for classifying are those that are unambiguous, obvious, and based on observations—not inferences.

> **FOR REFLECTION AND DISCUSSION**
>
> In the game Guess My Rule, one person takes a set of objects and divides the objects into two groups according to a feature that is observable. Someone else is challenged to determine the rule that was used to sort the objects. For example if we regarded the letters on this page as objects, we might sort them into those with a tail (e.g., q, p, and g) versus those without a tail (e.g., a, s, and m). Try a few rounds of Guess My Rule with someone else. What criteria did you use to avoid using inferences to divide the objects?

Dividing into Two Subgroups

A second guideline for classifying involves the number of groups created with each property. A property that follows Guideline #2 should divide objects into only two groups and should be specific enough to avoid the danger of some objects being classified in both groups. Guideline #2 insists that we not use a single property to subdivide a group into more than two subgroups. To accomplish this, we use a rule for separating the objects that is clear and does not require us to make a judgment call. You can think of classifying as a process of walking down a path with many branches. At each branch you can go to the left or to the right—you just have to choose one or the other. In our classifying path we never come to a fork in the road that has three possible routes.

It seems that a Venn diagram would be useful for classifying. Venn diagrams are visual guides for organizing information and are used in elementary and middle school classrooms (Moore, 2003). Venn diagrams provide teachers with another tool for clarifying their instructions and students with a guide for organizing their thinking. Typically Venn diagrams consist of circles to show the relationships between different objects. Figure 3.2 shows three possible arrangements of Venn diagrams for representing two sets. However, in the process skill of classifying, only one of these Venn diagrams is appropriate. The Venn diagram to the left sorts objects into one of two alternate groups, ensuring

 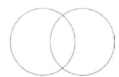

FIGURE 3.2: According to Guideline #2, only the left Venn diagram represents how objects are sorted into two groups.

that an object will always be classified in one category or the other. The other two Venn diagrams don't allow for the either/or approach that is required by the second guideline.

Imagine that you dump the contents of your book bag onto a table. In trying to classify these objects, you could use countless properties. In keeping with Guideline #1 you should sort them according to properties such as color or label. But you shouldn't use a possibly undetectable property such as cost (unless everything still carries a price tag) or origin (unless a label designates where the object was made)—sorting by these properties might require inferring. For students who struggle with using Guideline #1, you might find it useful to tell them to pretend to not know what the objects are or how they are typically used. Therefore students will be more apt to focus on the physical properties of the objects rather than on their purpose. If you are uncertain whether a property is effective as a classifier, consider the criteria we used for judging an observation. Properties that are based on factual observations and don't require any judgment or opinion are the best.

Our goal becomes identifying the properties that clearly divide the contents of your book bag into one group or the other. If a quality such as color is used, it is better to divide objects according to one color (i.e., whether it is a specific color) rather than a variety of colors. Recall that Guideline #2 is about dividing into one of two categories. But you need to make some distinctions to make the either/or practice clear. If you choose to classify objects according to color, then you probably should use a flexible system that dictates something like "contains some yellow" versus "contains no yellow."

Size can also be an effective property for classifying objects, but your rules for doing so should be as clear as those we just discussed for separating objects according to color. For example dividing objects into two groups of "big" and "small" seems to fulfill Guideline #2—except that it creates problems with Guideline #1. Is your textbook big or small? It's big compared to a car key but not compared to a car. To avoid such problems, you should standardize the classifying property size. For example you could divide objects relative to the size of your hand. In this case, a book is big, whereas a car key is small. Alternatively, the property of size could be based on an actual measurement (e.g., "taller than 15 centimeters" or "less than 100 grams"). Such measurable properties are not open to debate (Guideline #1) and clearly segregate the object into one of two opposite categories (Guideline #2). Now it's your chance to practice both of these.

Figure 3.3 shows an overlapping Venn diagram. We labeled the left circle "white shapes" and the right circle "square." The labels communicate observable properties and seem to be OK according to Guideline #1. However, some objects in the diagram could be sorted into both circles, violating Guideline #2. So you will need to identify a new property according to which to divide these thirteen objects. A good way to start is to redraw this Venn diagram so there are only two circles and they do not overlap at all. There could easily be a dozen different ways to classify these objects in ways that meet the criteria for both classifying guidelines.

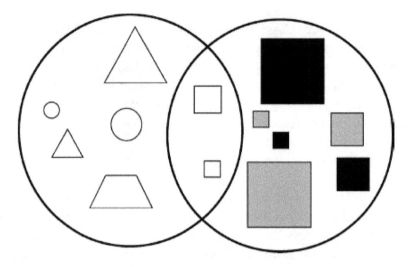

FIGURE 3.3:   This classification system fulfills Guideline #1 but not Guideline #2.

The Endpoint of Classifying

Guideline #3 states that the classification process is complete when each object is in its own group. Starting with four objects, we must first divide them into two groups and then further divide the subgroups by other criteria. The property that we use to sort the objects might initially divide them into groups of two and two or three and one. The goal of a complete classification system is to continue the sorting until every object is by itself. For the shapes exercise in Figure 3.3, this would require dividing by a series of properties until each object is in a separate category. How many times would the objects need to be divided? The answer is the number of objects you have less one. For the thirteen shapes shown in the figure, you would need to divide and subdivide the objects twelve times for each object to be by itself.

If you follow the three guidelines, you will find classifying to be a very powerful tool for organizing and thinking about a set of objects. Creating a classification system for a collection of leaves, stones, or kitchen utensils requires close observing of details. Along the way, patterns that were previously unnoticed may appear. Classifying is a process even very young children can use. Although we would not expect kindergartners to be pushed to follow Guideline #3, we could reasonably expect them to learn about Guideline #1 and perhaps Guideline #2 by having them engage in a game of "Guess My Rule" using a jar of buttons. In summary, classifying dictates that one must divide objects into clear and observable properties.

POINT-COUNTERPOINT

The use of science process skills with science instruction has been a common theme for more than four decades of science teaching and learning. However, process skills do not seem to have achieved universal application in American elementary and middle school science classrooms. In response, some might suggest that the push for process skills should be increased. Yet others question whether process skills are actually that appropriate in every situation. So in response to the question "What are the benefits of the science process skills?" we consider two differing opinions.

Science Process Skills: Tools for Everyone to Become Lifelong Learners (and Lovers) of Science

Cynthia A. Lundeen
Science Teacher Educator at Florida State University and Former Elementary School Teacher

Educating young children in science is a particularly "slippery slope" for teachers who often must negotiate time and talent for teaching this often elusive subject. Content or process? Seeing or doing? Best practice or least path of resistance? Hands on or off? Clean classroom or messy classroom? If truth be told, science process skills are largely underrepresented in many science lessons and classrooms, despite abundant research supporting their benefits.

Why this unfortunate fact? Attitude! Just say the word *science* in an open forum and watch people recoil. Many people view science as an intimidating mass of facts, figures, symbols, and substances beyond comprehension. This pervasive view of science as an encyclopedia of content just may be stunting our national growth in science achievement. Raising yet another generation of "science haters" reflects not on the product but on the teaching processes responsible for yielding poor achievement and poor attitudes. Attitudes toward a subject can be as important as the subject itself! Far too many teachers approach science teaching as a body of knowledge to be memorized (and scorned!) by students. Without attaching a meaningful connection to what is learned, students easily lose ground in learning science content and concepts. Science content meaning comes from connecting action, or science processes, to concepts.

So what is the value of the science process skills in teaching and learning science? To provide tools for lifelong learning! Changing attitudes and achievements in science education demands action! In early through middle childhood, *science* should be viewed as a verb as opposed to a noun. Introduced and practiced as a way of thinking and acting, "learning by doing" science yields a more meaningful understanding while promoting positive attitudes about science. I believe that one of the most significant goals of meaningful learning is teaching students, all students, to think and problem solve through active participation and application. Science teaching can contribute to this goal by maximizing opportunities to manipulate the physical world through the science process skills. These skills can enable all students to process new information through concrete experiences by sharpening thinking skills and building new understandings.

The notion of teaching science through an emphasis on process skills is not new. According to the National Science Teachers Association (NSTA), a "reasonable portion" of a science curriculum should emphasize science process skills. We know that in early childhood years, children begin to build understandings and beliefs about the world around them through physical manipulation and personal experiences and interactions as they continuously attempt to build more complex understandings of their world. Capitalizing on this critical learning time means teachers need to link science concepts to children's lives through methods uniquely associated with the process skills. In addition, young students develop a healthy skepticism about their environment as they grow, integrate, and actively negotiate their world. Teaching and learning through the science process skills maximizes the potential of every child to learn (and love) science!

Just How Universal Are Science Process Skills?

Malcolm Butler
Science Teacher Educator at the University of South Florida and Former High School Science Teacher

One day while sitting in my office preparing a lesson on the basic science process skills for a group of twenty-five young ladies soon to become elementary teachers, my mind began to wonder (as it often does). This time, I wondered about a possible connection between what we had done in a previous class on working with students from diverse backgrounds and the use of science process skills for understanding the natural world. Since the days of BSCS (Biological Sciences Curriculum Study) and SAPA (Science: A Process Approach) elementary science curriculum in the 1960s, all elementary school science teachers have been inculcated to believe that science is a set not just of facts but also of skills and scientific habits of mind. But lately I've begun to wonder … .

The process skills are considered universal. Along with the universality of the process skills is the commonly accepted idea that all children can learn. Inherent in this idea is that teachers must be able to teach students from diverse backgrounds, teaching them science in a way that is meaningful to them. There are several resources available to assist teachers in connecting the process skills and students' cultural background. For example Bazin and Tamez (2002) and the Exploratorium Teacher Institute have a set of activities that address the process skills from a multicultural perspective. So to think that there could be some aspects of the process skills that are inconsistent with some students' way of thinking and seeing the world could be heresy. Could it be comparable to being a member of the Flat Earth Society (Russell, 1997)? Could I be hanged for considering that there are some cultural issues that the teaching of process skills doesn't necessarily address?

But wait a minute. What about the kid who is struggling with the idea of grouping and classifying? Could it be that for this student, the idea of putting items in groups according to some characteristic goes against his home culture's belief that we should see the oneness of the world and all things? To take it one step further, consider that a dichotomous key proposes that either an item has or does not have some property, which forces one to have a maximum of two groups. Could this idea of separating one from many be difficult for students who think monolithically?

One more example about the possible disconnect between culture and the process skills comes to mind. Recently I was observing a group of first-grade students inquiring into magnets. As a diverse group of four students tried to figure out how magnets work, one of them suggested that the group try to test the strength of the magnet by measuring its length. Her reason for suggesting the idea was that her family had taught her that the taller a person is the stronger he or she is. Height is a sign of authority in this child's family. Now how would most teachers deal with this girl's train of thought? I think we would probably take the student's preconception (indeed, misconception) and move the student away from the idea that length equals strength and authority to the more accurate conception, without asking the question of why. So even though the child may come away with an understanding of what determines a magnet's strength, will she internalize the new concept or will she continue to hold onto a tenet of her family life?

It seems that we must admit that the process skills are a Western phenomenon. Although the Western origins and universal acceptance of process skills do not eliminate them from universal utility, we need to recognize that they can be more difficult for some students to grasp because they are incongruent with the students' cultures. Although in the past we've viewed process skills as this universal pathway for all students to move into science, I suggest that teachers need to consider the potential cultural baggage that travels with the process skills.

Oh well, enough wondering for this day. Maybe tomorrow I'll take a closer look at the geocentric view of the universe and the learning cycle ... .

CREATING VISIBLE CLASSIFICATION SYSTEMS

When the classifying of a group of objects is completed, then the outcome needs to somehow be recorded. From a teaching perspective, this is necessary so you can assess how well each of your students understands how to classify. In addition, translating ideas about how to sort objects into a written or inscribed form reinforces and clarifies understanding in the mind of the person who is developing the classification system. Finally, once a classification system is completed, it can serve as a tool for others. Having it recorded on paper makes it possible for others to apply the system to objects with which they are unfamiliar.

Classifying with Dichotomous Keys

The most common method of creating a visible classification system in science is the **dichotomous key**. A key is a tool for identifying something that is unknown—think of it as an answer key. The word *dichotomous* simply means that the key is based on the rule of dividing, and subdividing, into only two groups. Dichotomous keys are very common in field guides. These reference materials are useful for helping us to identify birds, trees, or rocks. The procedure is pretty straightforward. You start with the first pair of descriptions and decide which statement applies to your unknown object. Then you follow the next step. If it says "Go to ... ," then you jump down the key to that place, even if that means skipping other paired statements. When you reach a dead end, then you have identified the object. Here's an example of a dichotomous key that can be used to identify dried beans.

Dichotomous Key for the Contents of a Soup Bean Mix

1a	Bean shape is round	**Garbanzo bean**
1b	Bean shape is not round (oblong)	Go to 2
2a	Bean is dark in color	Go to 3
2b	Bean is not dark in color	Go to 4
3a	Bean color is solid	**Kidney bean**
3b	Bean color is speckled	**Pinto bean**
4a	Bean is entirely white	**Navy bean**
4b	Bean has a dark spot	**Black-eyed pea**

With a garbanzo bean, sometimes known as a chickpea, in front of you, you could possibly identify it immediately by reading the first line of this dichotomous key. Instead imagine you had a bean-shaped object that was brownish with dark spots. You would use the dichotomous key by reading the first statement pair and deciding which of the two statements described the object. Because of the bean's shape, line 1b is the better description, and it directs you to statement pair 2. Here you are to focus on the color. Because the bean is dark, you are sent to statement pair 3. Between those two statements, line 3b describes your object, which you now know is a pinto bean. Had you tried to identify a black-eyed pea, you would, as always, have begun at statement pair 1, which would have sent you to pair 2 because of the bean's shape. Here you would be directed to "go to 4," so you'd skip over and ignore pair 3. Because line 4b is the better description, you would now confirm the identity of your bean.

Looking back through the bean dichotomous key we can recognize the existence of all three classifying guidelines. First of all, because all properties are observable (color and shape), it is unlikely that any inferring would be required. This is consistent with Guideline #1. Next, this system follows Guideline #2, because every set of properties is an either/or situation, as shown by the statements always being in pairs. Last of all, if we assume that we began with five types of beans, then each object would result in its own category, which is in keeping with Guideline #3. Notice that to completely classify all five beans, our key needed to sort the objects by a number of properties equal to one fewer than the number of objects. Now if you were going to have students classify jelly beans, as Dave Crowther (2003) described in a *Science and Children* article (see the Suggested Readings section near the end of this chapter), and you started with twenty different types, students would need to create nineteen statement pairs to develop a complete dichotomous key. Maybe this will reinforce why it may be best to begin with a much smaller group of objects.

Classifying with Tree Diagrams

Another way to produce a visible classification system is to create a tree diagram. Although it looks very different from the dichotomous key, the way it functions and its adherence to the three guidelines are consistent. The key components of a tree diagram are an oval and two labeled arrows, such as what you see in Figure 3.4. The oval specifies the property you want to focus on—ignoring all possible others. Refer back to the first line of the bean dichotomous key. What property goes into the oval? Yes, it is shape. This alerts the person who is trying to identify a bean to which feature

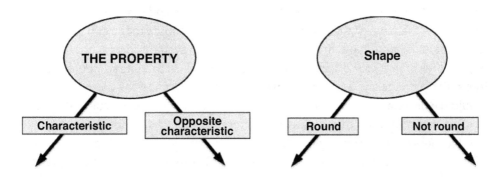

FIGURE 3.4: The basic unit of a tree classification diagram identifies the property and two opposite descriptions.

to pay specific attention. Shape, rather than color, size, or anything else, has been identified as the distinguishing characteristic.

The arrows are used to indicate which pathway to follow. Notice that there are only two pathways, never more than that. Because the choice is between only two pathways, it is clearest to label the arrows with characteristics that are the opposite of each other. Again using the bean dichotomous key as a reference, we would label one arrow "round" and the other "not round." Using "oblong" as the second characteristic might not be clear enough. We have now translated the first line of the bean dichotomous key into a basic tree diagram.

A tree diagram is a graphic representation of the first statement pair in a dichotomous key. Some people are more comfortable using the tree diagram as a classification system because it is so visual. The benefit of the dichotomous key is that it takes up less space on the page. You may have a preference for one over the other, and that's fine. Both rely on the same classifying thought processes and are consistent with the three classifying guidelines. You should be able to understand both ways of representing the completed classification system and be able to move back and forth between the two without much difficulty.

Figure 3.5 represents a complete tree diagram that relies on the same properties and characteristics as the dichotomous key for the beans. You can double-check the tree diagram to compare its consistency with the dichotomous key. We can imagine that the ovals in the tree diagram could be labeled with the line numbers from the dichotomous key. Meanwhile, the arrows would be identified with "a" or "b." Using either system, that is, the original dichotomous key or this tree diagram, someone would be able to identify a dark red, bean-shaped object as a kidney bean and confirm that a spotless white bean is a navy bean.

CLASSIFYING AND CULTURAL NORMS

The process skill of classifying can be challenging, because it's much more complicated than making an observation or creating an inference. There are also cultural assumptions that are tucked within the three guidelines and that, for those who hold a worldview that doesn't mesh with the guidelines, contribute to the difficulty of classifying. Sometimes students struggle with classification systems for reasons that extend beyond their inability to comprehend the goal. Even adult students sometimes express frustration with making a classification system. It's almost as if the thought processes required when classifying, which is obviously based in the science culture, are a way of examining the world that is unusual for some people. As a consequence, we find that teachers may very much need to treat classification, particularly as it is a process of science, as a foreign culture. Teachers may need to give additional guidance to some students. An effective way teachers can provide this guidance is to move those students back and forth between their ways of thinking and the science culture's ways.

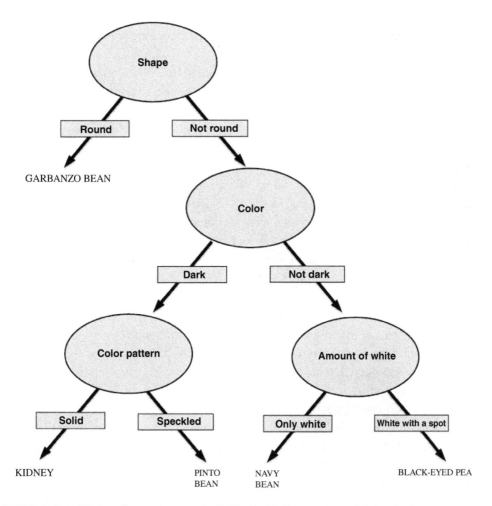

FIGURE 3.5: This tree diagram is conceptually identical to the soup bean dichotomous key.

Modern science emerged out of traditions that are clearly Western. It seems that the process of sorting into either/or categories can be traced to ancient Greek thought. The process of organizing objects according to whether they have or do not have a particular property, with no intermediate category, is sometimes called Aristotelian (Bowker & Star, 1999). This very formal approach to classifying, unlike the version of classifying we might use in everyday living, makes no allowance for fuzziness. It is important to recognize that thinking scientifically is not natural or automatic. Indeed learning to perceive the world in ways consistent with this dichotomous, observation-based perspective is something we must learn. For some of us, this may seem to be very natural. In actuality it is an extension of the cultural traditions within which we were raised. Classifying is an example of a very particular way of thinking that teachers must introduce to students to familiarize them with the culture of science.

Here is a very specific example comparing people from two cultures. Researchers presented study participants with proverbs that seemed to contradict themselves, such as "Beware of your friends, not your enemies" (Peng & Nisbett, 1999, p. 744). This proverb suggests we should be cautious about the people we have grown to trust. Such contradictory proverbs, along with non-contradictory proverbs, were presented to college students in the United States and to similar-age students in Taiwan. The American students showed a dislike for contradictory proverbs while the Chinese students preferred those. This suggests that a Western view favors ideas that are internally consistent while an Eastern view is comfortable with such ambiguity. The bottom line is that Western thinking favors either/or categories while Eastern thinking accepts less clear-cut divisions. This illustrates how cultural background may complicate the classification guidelines for some students, and make the guidelines more difficult for such students to grasp.

A challenge we have as science teachers is finding the right balance between the culture of science and the cultures of the students. Even though many individuals have made contributions to science, the culture of scientific thinking is, for better and for worse, not an approach that makes sense to all cultures. This becomes an opportunity to reinforce that culture is much more than simply a student's family life. In addressing the broad challenges of classifying, researchers have recognized that learning to classify is complicated by culturally bound ways of thinking: "Categories are learned as part of people's membership in communities of practice in that categories are tied to each community's particular usages and practical requirements" (Vosniadou, Pagondiotis, & Deliyianni, 2005, p. 118). The implication in the science classroom is that learning to formally classify is not simply something students must absorb. Instead classifying in a scientific fashion is part of a very specialized culture that can be odd to those who have not been enculturated into either/or mindsets.

The culture of science reveals itself in the classification guidelines. First, science places a great deal of emphasis on facts—little tolerance is given to knowledge claims that can't be supported by data. When a team of scientists claims to have made a new discovery, no one pays much attention unless the scientists are able to provide data to support their claims. Even then, other scientists will conduct similar experiments to see whether the same data can be gathered. This is apparent within classifying Guideline #1, which makes no allowance for inferring within a system of classification. Second, the actions of science leave little room for shades of gray. Thinking scientifically requires a worldview that is largely dichotomous. These cultural norms of science (i.e., the demand for facts and the either/or mentality) can be in conflict with other ways of thinking about the world. This competitive spirit, the appeal for either/or thinking, which seems so much a part of American culture, is very much at odds with many "non-Western" cultures. In this regard, creating a classification system may be unnatural for individuals who don't view their world in such a competitive fashion.

How do we negotiate such a cultural conflict? What can a teacher do when students appear to struggle with classifying and are seemingly resistant to the process? One approach is to believe that some students can learn how to classify while others cannot—and we are NOT recommending this tactic. A far better approach is to think about the situation as an opportunity to help a person from one culture learn about another culture. This means making the cultural norms, the ways we think in a culture, very explicit.

Imagine science as a special cultural neighborhood in your community—a place where students may not have visited. Your goal is not to make them sacrifice their own culture but instead to make them aware of and appreciate a different culture. In preparation for your "science-town" field trip, you explain to the students how to act so they aren't perceived by the locals as being disrespectful. You need to tell them what is appropriate, perhaps even writing down the behavioral norms and posting them around the room for all to read. Learning how to classify is similar to a cross-cultural field trip. As students present the classification systems they've developed, you should explain the norms again, referring to the charts posted around the room, and help students understand the underlying reasons for these norms. This disposition toward teaching children how to classify—how to move toward this scientific action—is much healthier than saying that "some kids don't think that way" and more likely to be effective.

FOR REFLECTION AND DISCUSSION

Consider the value of balancing the content of science with the process skills of science. What would it be like if an elementary school science curriculum emphasized content to the exclusion of process? In contrast, what would be some drawbacks of a science curriculum that focused only on process skills and spent little energy on teaching children science concepts?

Process Skills and Students With Cognitive Limitations

Just as the actions of science may be unfamiliar to students from non-Western backgrounds, so too are these actions difficult for students with cognitive limitations. It is important to recognize the cultural backgrounds and linguistic capabilities our students bring to school, and it is equally essential to become familiar with the cognitive abilities students carry with them. For some students, process skills such as classification represent a substantial intellectual challenge. As teachers we need to understand the level of cognitive challenge that is appropriate for the students in our classroom. This understanding should be based on an assortment of information sources.

Credible sources of data include insights from the parents, discussions with special education teachers, results of past testing, and our observations of student performance. However, the inferences we make about students' cognitive abilities must be understood as tentative. This is because a student's participation in the actions of science can help them move above their current capabilities. If a student enters our room at the beginning of the year having great difficulties with the prospects of creating a classification system, this should not suggest that he or she can never have success with this process skill.

The desire to help all students develop their science process skills is crucial, but desire, hope, and good intentions are not enough. As teachers of science, we must know the actual strategies we should employ to help all students achieve their highest potential. This involves not only giving the students time and opportunity to practice and refine their use of the process skills but also appreciating the emotional, social, and behavioral dimensions of classroom tasks. Teaching the science process skills places demands on students that extend beyond the cognitive aspects of observing, inferring, and classifying.

Maroney, Finson, Beaver, and Jensen (2003) suggested that helping students, particularly students with learning disabilities, to succeed in classroom science should begin with the teacher's careful analysis of the various skills required for completing a science activity. Within the context of a science activity, teachers should consider these categories of performance: classroom behavior, social skills, group coping, academic skills, science process skills, and scientific inquiry. In advance of a science activity, the teacher should closely analyze the various kinds of skills involved in an activity to decide the relative difficulty of the various skills and to determine if the activity is appropriate for all the students. If it is not, then the teacher can use the chart presented as Appendix A to inform modifications of the activity for some of the learners in the classroom. (See Appendix A for an example of a rubric to aid in this analysis.)

Then as the students work on an activity, the teacher can gather information about particular students in terms of their abilities on each of the categories of performance. By looking at these charts over time, teachers then have a sense of the students' development. What is as important is that such analysis of the demands of a task can also remind the teacher that teaching science process skills cannot be separated from the other demands students face, especially those students whose cognitive abilities are below the norm for their age group.

When carefully selected activities are employed, that is, those that challenge students' abilities in one or two of the categories of performance, but not all four, students can develop their competence. If the process skill to be developed is classification, students should be initially asked to observe and make inferences about a variety of materials, such as an assortment of buttons, shells, or screws. This can be followed by having students examine and use others' classification systems and creating and refining their own classification systems. As students make progress with the activities, teachers should limit other aspects that special education students might find difficult, such as allowing those students to work alone if social skills and group coping skills are particularly challenging. The challenge for teachers is to be aware of the challenge they provide for students and to make conscious decisions about the number of challenges a student can be expected to meet. Through a progression of activities, all students can become proficient with process skills, such as classifying in a genuinely scientific fashion—even though it might take some students a little more time to reach that point. Growth and improvement should be the goal for each of our students, particularly for those students challenged by special cognitive limitations.

As teachers of science, not only must we recognize that we need to give the students time and opportunity to practice and refine their use of the process skills but also we have to appreciate the emotional, social, and behavioral dimensions of classroom tasks. One of the reasons that science can be so powerful for students with cognitive limitations is that the subject can provide the context that is so necessary for learning. Process skills supply this context because the students are in direct contact with the subject matter (Patton, 1995).

SCIENTISTS' USE OF PROCESS SKILLS

A reasonable question about the science process skills is whether emphasizing their use by elementary and middle school students represents authentic science. Are professional scientists really involved in the processes of observing, inferring, and classifying? Or are process skills an oversimplified attempt to have children engage in scientific activities? A rare and revealing peek into the thought processes of a scientist can reveal the obvious value of process skills such as observing and inferring.

Richard Feynman won a Nobel Prize for his work on quantum electrodynamics, and even if we wanted to explain this field of science to you, we really can't. We thought it might be interesting to have a glimpse into his mind. Rather than try to understand his ideas regarding photons and matter, we'll draw on an earlier example of his investigative thinking. Writing about his life as a scientist, Feynman described, in the following passage, his curiosity about ant behavior:

> One question that I wondered about was why the ant trails looked so straight and nice. The ants look as if they know what they're doing, as if they have a good sense of geometry ... .
>
> The moment the ant found the sugar, I picked up a colored pencil I had ready (I had previously done experiments indicating that the ants don't give a damn about pencil marks—they walk right over them—so I knew it wouldn't disturb anything), and behind where the ant went I drew a line so I could tell where his trail was. The ant wandered a little bit wrong to get back to the hole, so the line was quite wiggly, unlike a typical ant trail.

When the next ant to find the sugar began to go back, I marked his trail with another color. (By the way, he followed the first ant's trail back, rather than his own incoming trail. My theory is that when an ant has found some food, he leaves a much stronger trail than when he's just wandering around.)

The second ant was in a great hurry and followed, pretty much, the original trail. But because he was going so fast he would go straight out, as if he were coasting, when the trail was wiggly. Often, as the ant was "coasting" he would find the trail again. Already it was apparent that the second ant's return was slightly straighter. With successive ants the same "improvement" of the trail hurriedly and carelessly "following" it occurred. I followed eight or ten ants with my pencil until their trails became a neat line.... It's something like sketching. You draw a lousy line at first; then you go over it a few times and it makes a nice line after a while. (Feynman, 1989, p. 79)

Here, in the words of one of the leading scientists of his era, we can witness the process skills being put to actual use. We notice how he uses pencils to assist with his observing. We also can detect his efforts to infer the reasons for the ants' behaviors. All in all, without any fancy scientific terminology, a scientist demonstrates his craft and does so in a way that makes the science process skills evident.

MORE THAN BASIC SCIENCE PROCESS SKILLS

Within this chapter we've devoted attention to the first three of six basic science process skills. In the next chapter, we will present the balance of the list: measuring, predicting, and communicating. In total, the six basic science process skills form an important basis for an elementary school science program. We can realistically expect students from kindergarten and the higher grades to actively engage in science through their use of these six skills (Padilla, 1990). However, as they mature, students will have the cognitive abilities to learn even more complex science process skills. We present them here only so you are aware of their existence. On the basis of our experiences, as well as our work with classroom teachers, we are very cautious about expecting students to be proficient at the integrated process skills before age eleven years or thereabouts.

The **integrated science process skills** build on the fundamentals provided by the basic process skills. What makes this new list of skills "integrated" is that to undertake them requires integrating or combining the basic science process skills (Rezba, Fiel, Funk, & Okey, 1994). For example to make a graph, which is one of the integrated science process skills, a student needs to be skilled at observing, measuring, classifying, and communicating. Making a graph involves combining all of these into an integrated whole. The reason younger children should not be expected to master the integrated science process skills is that, typically, their minds have difficulty coordinating the thinking required to perform these tasks. It's almost as if younger brains don't have the capacity to manage the integrated skills—much like when a toddler attempts to carry more toys than he or she is capable of holding. At best, children in the primary grades require a great deal of support and guidance by their teachers to master an integrated science process skill. Before children are eleven years old or so, it is probably not worth all of the effort (Kuhn & Dean, 2005). But by the time children are in fifth or sixth grade, their minds seem to be able to handle the complexity required to use the integrated process skills. Our stance is that the best way to approach this is to look at the big picture: the integrated science process skills only make sense within the context of designing and conducting scientific experiments.

Experimenting requires using all of the basic and integrated process skills, much as one would use for a science fair project. Even though many people use *experimenting* as a synonym for any science activity, such as when a child asks, "Are we doing experiments today?" we view experimenting as much more involved and ambitious. Our preference is to use *investigating* as our catchall phrase to describe science activities. When it comes to experimenting, that involves much greater complexity than can be captured in just one section of a textbook. Therefore, we have devoted all of Chapter 5 to this topic.

CHAPTER SUMMARY

- The science process skills supply us with the actions that complement the concepts of science. The process skills describe the things students should do in science, not just experience by hearing about others doing science. Although becoming skilled users of the science process skills helps students gain entry into the culture of science, they also support the development of language fluency and contribute to the development of a coherent classroom community.

- Observing and inferring are intermingled actions: observing involves gathering facts, whereas inferring attempts to explain those facts. Considering how closely the statement is to being an indisputable fact allows us to assess the quality of an observation.

- Inferring involves posing a statement that is intended to explain observations. Inferences are evaluated by considering the strength with which they account for what has been observed.

- Classifying allows scientists and students to sort through observations and organize them by following specific guidelines. Key guidelines include the absence of opinions and the reliance on categories that sort objects into one of two opposite subgroups.

- Science process skills are a cultural norm of science. These norms may conflict with other cultures' views of the world. In particular, the either/or and bias-free properties that are used within classification are representative of a certain worldview that may not be self-evident to those who are familiar with contrasting perspectives.

- Experimenting is considered an integrated science process skill and involves a collection of skills normally not accessible to students until they are in the upper elementary grades. The six basic process skills can be used, to varying degrees of complexity, from kindergarten through middle school.

KEY TERMS

Classifying: a basic science process skill involving the organization of objects into a comprehensible framework based on observable properties of those objects.

Dichotomous key: a common tool of classification that allows the user to identify some item in the physical world. (Many field guides of rocks, animals, and plants are based on dichotomous keys.) Such keys are constructed by dividing all groups of items into one of two parts, so that the use of keys consists of a series of choices, one choice for each step, leading to the proper identification of the item.

Inferring: one of the essential science process skills, involving generating explanations based on valid observations. In the culture of science, better inferences (the explanation generated through the process of inferring) are always considered possible.

Integrated science process skills: the complex actions appropriate for use starting in the upper elementary grades. The skills included in this list are defining operationally, controlling variables, stating hypotheses, reading and making graphs, and designing experiments.

Objectivity: a characteristic of scientific actions in which personal biases are minimized, often through the process of limiting the role of inference in the collection of scientific observations, allowing for a more straightforward portrait of the physical world.

Observing: one of the essential science process skills, involving the active use of the senses to make direct descriptions of some aspect of the physical world.

Science process skills: those discrete but essential actions of science that are used to conduct scientific inquiries. Process skills include observing, inferring, classifying, measuring, predicting, and communicating.

SUGGESTED READINGS

Wood, J. (2005, April/May). Discovery central. *Science and Children, 42,* 36–37.
> Rural teacher Jaimee Wood explains how centers are used in a rural classroom. Even though the students are very young, Wood has kindergarten students rely on all the science process skills during a unit involving plants.

Crowther, D. T. (2003, October). Harry Potter and the dichotomous key. *Science and Children, 40,* 18–23.

Dave Crowther teaches elementary science methods in Reno, Nevada, and he created a clever way to teach classification. He relies on a learning cycle lesson plan, something we will address later in this book, to help children classify jelly beans. More than just cleverly using candy, Crowther illustrates very effective planning and a memorable approach to making dichotomous keys.

REFERENCES

Amaral, O. M., Garrison, L., & Klentschy, M. (2002). Helping English learners increase achievement through inquiry-based science instruction. *Bilingual Research Journal, 26,* 213–239.

Bazin, M., & Tamez, M. (2002). *Math and science across cultures: Activities and investigations from Exploratorium Teacher Institute.* New York: New Press.

Bowker, G. C., & Star, S. L. (1999). *Sorting things out: Classification and its consequences.* Cambridge, MA: MIT Press.

Dawes, L. (2004). Talk and learning in classroom science. *International Journal of Science Education, 26,* 677–695.

Dewey, J. (1991). *How we think.* New York: Prometheus. (Original work published 1910.)

Doyle, A. C. (2004). *The adventures and the memoirs of Sherlock Holmes.* New York: Sterling Publishing.

Duffy, G. G. (1998, June). Teaching and the balancing of round stones. *Phi Delta Kappan, 79*(10), 777–780.

Feynman, R. (1989). *Surely you're joking, Mr. Feynman: Adventures of a curious character.* New York: Bantam.

Harlen, W. (2000). *The teaching of science in primary schools.* London: David Fulton.

Kuhn, D., & Dean, D. (2005). Is developing scientific thinking all about learning to control variables? *Psychological Science, 16,* 866–870.

Linik, J. R. (2004). Growing language through science. *Northwest Teacher, 5*(1), 6–9.

Maatta, D., Dobb, F., & Ostlund, K. (2006). Strategies for teaching science to English language learners. In A. K. Fatham & D. T. Crowther (Eds.), *Science for English language learners* (pp. 37–59). Arlington, VA: National Science Teachers Association.

Maroney, S. A., Finson, K. D., Beaver, J. B., & Jensen, M. M. (2003). Preparing for successful inquiry in inclusive science classrooms. *Teaching Exceptional Children, 36*(1), 18–25.

Menand, L. (2001). *The metaphysical club: A story of ideas in America.* New York: Farrar, Strauss and Giroux.

Mill, J. S. (1884). *A system of logic, ratiocinative and inductive; being a connected view of the principles of evidence, and the methods of scientific investigation.* London: Longmans, Green & Co.

Moore, J. E. (2003). The art of sorting: Using Venn diagrams to learn science process skills. *Science Activities, 39*(4), 17–21.

North Central Regional Educational Laboratory. (2004). *All students reaching the top: Strategies for closing academic achievement gaps*. Naperville, IL: Learning Point Associates.

Padilla, M. J. (1990). *Research matters to the science teacher: The science process skills*. Retrieved from http://narst.org/publications/research/skill.cfm

Patton, J. R. (1995). Teaching science to students with special needs. *Teaching Exceptional Children, 27,* 4–6.

Peng, K., & Nisbett, R. E. (1999). Culture, dialectics, and reasoning about contradiction. *American Psychologist, 54,* 741–754.

Rezba, R. J., Fiel, R. L., Funk, H. J., & Okey, J. R. (1994). *Learning and assessing science process skills*. Dubuque, IA: Kendall-Hunt.

Russell, J. B. (1997). *Inventing the flat earth: Columbus and modern historians*. Westport, CT: Praeger.

Scudder, S. H. (1879). The student, the fish, and Agassiz. In *American Poems* (3rd ed.). Boston: Houghton, Osgood and Co.

Shaler, N. S. (1946). The autobiography of Nathaniel Southgate Shaler. In H. Peterson (Ed.), *Great Teachers* (pp. 213–215). New York: Vintage Books.

Vosniadou, S., Pagondiotis, C., & Deliyianni, M. (2005). From the pragmatics of classification systems to the metaphysics of concepts. *Journal of the Learning Sciences, 14*(1), 115–125.

Weaver, H. N. (2000). Balancing culture and professional education: American Indians/Alaska Natives and the helping professions. *Journal of American Indian Education, 39,* 1–18.

READING 4

THE 5E LEARNING CYCLE AS A MODEL FOR SCIENCE TEACHING

By John Settlage and Sherry Southerland

HIGHLIGHTS

- Teaching students to learn any subject involves helping them grasp the bits of information and the broad ideas. Planning to teach requires careful decision making about the sequence of the parts and wholes.

- When teaching a topic starts with the main ideas and then moves to specifics, it is labeled *deductive*. In contrast, when teaching begins with an analysis of the parts and subsequently moves toward the whole, it is called *inductive*.

- Teacher support of student learning is gradually reduced as confidence and competence increases. Learning should ultimately result in students having independent control over their work.

- Learning cycle instruction begins with hands-on activities and then moves to the formal presentation of concepts. Following these two stages within the learning cycle, students apply their new ideas to fresh situations.

- Every variation of the learning cycle contains the same key elements. In this text, the model consists of five phases: Engage, Explore, Explain, Extend, and Evaluate.

- Students with a wide variety of backgrounds can learn science through the learning cycle. Background experiences are the starting point that connects every learner to the content.

- Students are expected to master one central science concept after passing through a single learning cycle. This allows students to apply their newly acquired ideas before the teacher moves on to the next concept.

In the preceding chapter we examined three approaches to science teaching. Each approach originated in response to a perceived challenge within science education. The discovery approach provided students with activity-centered lessons, inquiry-based teaching allowed students to participate in the thoughts and actions of science, and conceptual change instruction took its cues from the need to ensure that teaching is actually contributing to students' learning of science concepts. However, it is our view that none of these options provides sufficient detail about how a teacher would actually implement any of these approaches in an elementary or middle school classroom as a central aspect of science instruction. Of even more importance in our view, however, is that none of the approaches pays particular attention to student diversity. Our aim within this chapter is to provide a model of science teaching that resolves these problems.

Within this chapter you will learn about a powerful model of teaching science called the learning cycle. This instructional model is a by-product of the science curriculum heyday of the 1960s and 1970s. Over the years this model has been revised to its current form that we will present to you. There is considerable research demonstrating that this instructional model is a powerful tool, not only for advancing the quality of science instruction but also for potentially fulfilling the goal of "science for all" without excluding any group of students. The phases of the learning cycle are interconnected and interdependent. A teacher who is using the learning cycle effectively has to pay attention to the central science concept and to the details that become important during each phase. Because this approach is so effective at helping children develop solid understandings of the objects of science (scientific explanations), it is important for teachers to understand the learning cycle model and learn to use it in ways that allow them to make it their own.

MODELS OF TEACHING

As you know from your time as a student, there are many ways in which a subject can be taught. The teacher might lecture, the students might work in groups, individual students might do independent projects, and so on. Each approach has its own assumptions and purposes. As someone matures into his or her role as a professional teacher, he or she should be masterful at using a variety of teaching models and know the underlying philosophy for each.

We can simplify teaching and learning by reducing it to two components. One component is the central idea that is to be learned. This is the big picture or one concept the teacher intends for the students to understand. The other component is the collection of examples, experiences, and evidence that supports the one concept. Activities that fall into this second component include book excerpts, video clips, websites, field trips, hands-on work, and so on.

One component of learning is the "whole," and the other component is what constitutes the "parts." For science learning to happen, a student needs to comprehend both components and be able to explain the ways those components relate to each other. In music the "whole" might be learning an entire song whereas the "parts" would include the melody, harmony, lyrics, and rhythm. Likewise a jigsaw puzzle can be understood in terms of both the individual pieces and the image that appears when all the pieces are properly assembled.

Imagine you are responsible for teaching a group of students with a variety of science and linguistic abilities about simple machines. You are provided with several models of simple machines, some basic equipment (pulleys and gears) in sufficient quantities for every student to investigate, a class set of trade books that describes simple machines in action, and a video that shows the use of simple machines in everyday life; these are the "parts" the students will experience. For the "whole" you have a teacher resource book that provides you with lots of background information about simple machines. You are given the freedom to design your own lessons to teach the students. Where might you begin?

Deductive and Inductive Approaches

One possible approach to teaching simple machines is to orient the students with what they will be studying. This includes providing a quick overview of simple machines. The teacher might explain that simple machines move objects by using a device that makes the work easier. All simple machines function under the same principle: to reduce the force required to move an object, the machine must exert effort over a longer distance. That's the one concept you intend for the students to learn. All of the equipment and other resources are then used over the next several science lessons (including hands-on activities, reading activities, discussions) to reinforce that concept in the minds of the students. An appealing aspect of this approach is the variety of materials that accommodate the various kinds of learners in the class. For the topic of simple machines, some students would respond to working with actual gears and pulleys, others would respond to animations and video clips of simple machines in action, whereas others would learn best by reading nonfiction texts.

In this scenario, what is the sequence of parts and wholes? The learning began with an emphasis on the global idea of simple machines followed by exposure to various examples. When teaching is organized in this way, that is, by starting with the whole and moving to the parts, it is called a **deductive approach** to instruction. Beginning with the general idea and then going to the specifics is a very common way of teaching, and it is probably the most common approach you experienced in your own science learning. A deductive way of teaching has many strengths. A deductive approach is an important tool within a teaching toolbox, especially when the big concept is very specific and can be reduced to a precise sequence.

However, the deductive approach is just one method of teaching (see Figure 4.1). For people who know only the deductive approach, their teaching toolbox is very limited. There is a saying that applies here: "To the person holding a hammer, everything looks like a nail." Obviously there are times when a hammer is exactly the right tool for a particular task, and when used properly a hammer can be an elegant and efficient tool. But in the construction trades there is never a person whose only job is hammering—knowing how and when to use other tools is important. As useful as deductive teaching is, we need other teaching tools to select from.

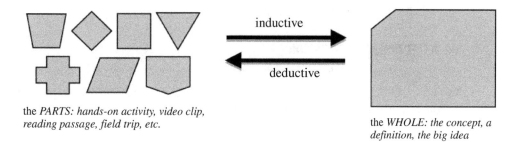

FIGURE 4.1: The relationship between parts and whole is how inductive teaching and deductive teaching are defined.

Another way to plan for teaching is the **inductive approach.** The key feature of an inductive approach is that learning begins with the parts and ultimately leads to the whole. The teacher gives the students experiences with many examples and activities without explaining how everything fits together. Eventually the teacher guides the experiences so the students come to understand how they all fit into a nice and complete whole.

To teach simple machines with an inductive approach, the lesson would begin with the activities. The teacher might have the students first spend a lesson or two working with ramps and inclined planes. They would use spring scales to measure the amount of force needed to pull objects up a ramp. Then the students might work with pulleys and explore what happens to the amount of force when multiple pulleys are used. Later they would explore the gears and how those work.

After all of these activities, the teacher would introduce the concept of simple machines. In explaining simple machines, the teacher would help students recognize how all their previous investigations share characteristics that are united under this big idea. The students have developed familiarity and comfort with the equipment, and they are probably finding some patterns in how the materials operate. The teacher's task is to help them to generalize from their varied activities.

We must acknowledge that this inductive approach flies in the face of this old-fashioned notion of teaching: "Tell them what you're going to teach them, tell them while you're teaching, and then tell them what you taught them." This doesn't mean that an inductive approach is a free-for-all. In fact, the teacher has to be especially thoughtful about the activities the students will do so they provide an adequate foundation for the big idea. No activity should be selected simply because the students will have fun. This would be as odd as having a jigsaw puzzle in which the manufacturer threw in some extra pieces just because they were pretty. In short, all of the parts should fit together to build the whole.

FOR REFLECTION AND DISCUSSION

In what ways have the authors used an inductive approach to introduce the concept of deductive? Similarly, how have the authors made use of a deductive approach to present the concept of inductive?

Student Benefits From Inductive Teaching

Because inductive teaching may not be an approach you've ever experienced, we'd like to give you some reasons to consider using it other than for its novelty. We also want to emphasize that inductive teaching isn't an all-purpose tool; it has its limits. But there is also a great deal of research supporting the use of inductive teaching within elementary school science instruction to support student learning.

In your teacher education courses, you may have heard that an effective teacher is one who starts where the children are. This means that for learning to occur, particularly in diverse classrooms, it must build upon what the students bring to the situation. No matter how young they are, what country they are from, or what language they speak, when children walk into a classroom they already hold many ideas about the world and the way that it works. Anyone who has spent more than a few moments with a child and believes that children are "blank slates" is not really paying attention to the child. You cannot and should not assume that everything you will teach is completely new to your students.

When we use *diverse*, we aren't using that term in a politically correct way to signify Black, low income, or urban. We don't mean to imply non-White either. Instead, *diverse* refers to a mixture of students who have different backgrounds, which include family income, ethnic heritage, skin color, physical ability, first language, mental capabilities, and so on. One startling discovery for far too many new teachers is the realization that the children in their classroom are not all the same. Instead they vary in many, many ways. It is important to recognize that these differences hold substantial implications for student learning and thus the appropriate teaching approach.

Our particular concern, and the motivation for developing this book, is the persistent science achievement gaps as reported in research and the popular press. From our perspective, these data clearly indicate that some students are in greater need of effective science teaching than others. We anticipate that schools that are more desirable because of sufficient resources, adequate facilities, and other factors known to support student learning will rarely have difficulty with recruiting teachers. However, students in less fortunate circumstances also require teachers, and these teachers must possess an especially potent mix of information and strategies—and commitments—to advance those students. In actuality, when we speak of diverse we have in mind those students who might otherwise not be adequately served by an educational system that is blind, indifferent, or simply incapable of responding to the challenges and opportunities represented by these children. Balanced against this is our firm understanding that it's really hard to know where your students are coming from in terms of their background experiences. In our highly mobile society, you may have students in your class who regularly visit relatives in another country, but you could also have students who have never traveled more than a couple of miles from their home. Some families may set aside one Saturday each month to visit local cultural sites (zoos, museums, concerts, and the like) while other families spend the weekends with their extended family. As a teacher of science you have this dilemma: you should begin with the students' experiences, yet they all have had different experiences! In many classrooms the students may have more in common with each other than with the teacher.

The simple fact is that having a teacher who does not live in the same neighborhood as the students can be a source of incongruities. You may not know about the annual summer street fair. You may not be aware of the city park where the children play in their free time, the stream that cuts through a local block of land, or the community garden that many of them frequent. You may not know about the after-school programs held at the community center. Without this knowledge, you will have a harder time coming up with examples from the students' lives to reinforce the concepts you are teaching them. What can you do? You understand teaching is more effective if you start with the students' previous experiences—but their lives are so varied that you're worried (and rightly so) about making unwarranted assumptions about what they do and do not know about the world. You may harbor the fear (again, rightly so) that you're going to disadvantage some students because you incorrectly assume they have experienced something that you want to use as the basis for an entire lesson. The reverse is also possible. If some of your students have limited English fluency, you may overlook the wealth of knowledge they bring from their lived experiences or schooling outside of the United States. In each of these cases, it is easy to misjudge students' background knowledge. Beyond missing opportunities to connect school science to your students' lives, you can reduce your effectiveness with responding to student behavior and create communication barriers between you and the families of your students.

The Power of Shared Experiences and Inductive Teaching

One solution is to involve all of the students in the same activity at the outset. By doing this, everybody has a shared experience they can all relate to. When the class begins to discuss their ideas, nobody is left out because they are unfamiliar with the scenario. This doesn't mean we neglect or ignore students' prior experience or knowledge. But by beginning with a common experience, we have less chance of some students making connections between the science lesson and their own lives and others being utterly confused because they cannot relate to the discussion. The value in this approach goes beyond just common experiences. For students learning English, such concrete and shared experiences are an excellent connection to which they can attach their growing English vocabulary.

Research has shown that students' reading comprehension and vocabulary development are increased when they are first provided with direct experiences. By beginning with such experiences and *following* with explicit treatment of the terms to be learned, students learn the vocabulary more quickly, tend to remember the vocabulary better, and are more likely to appropriately use the vocabulary in their writing and speaking. Likewise for students with learning disabilities, such initial, concrete experiences are incredibly important in allowing them not only to understand the objects of science but also to be able to retrieve those ideas when needed from their memory. The point we are making is that even though common sense might suggest that students need to be taught terminology first, a host of research indicates that the strongest foundation for learning science is provided by concrete experiences followed by formal vocabulary instruction. We recognize that this approach feels different, as most of us didn't experience science in this manner. But given what we know about how students learn, it is clear that experience followed by analysis of concepts is the most effective way to teach and learn science.

Teachers and those who are preparing to become teachers should be cautious about treating their childhood experiences as ones that all children have. Yes, each of us should dearly hold on to our family traditions and special events. But thinking that our personal experience is normal leads to the danger of our treating others' experiences as not normal. Even something as basic as tending a vegetable garden or swimming in a backyard pool may not be an experience you can assume all children have had. This may be true not only for students living in the city but also for students who live in prestigious suburban communities. In addition, many people take advantage of the fact that teaching positions are available in many parts of the nation. But after moving to a new community, teachers are often surprised that the local culture can be just as different as are the changes in climate or topography. As a result, what was normal in one setting can prove to be an unhelpful reference point in a new place.

Instead of assuming that every child has had a chance to plant seeds, mix substances, or take care of animals, it would be better to make those experiences the starting points for science instruction. Communities and cultures are built around shared experiences, and if we think of a classroom as a learning community, then providing opportunities for the members to have an experience they all share has great unifying potential. From a socialization perspective, shared experiences are important and should be a regular aspect of science. Certainly, having a shared experience is no panacea for eliminating differences in students' backgrounds and conceptual lenses. However, if the explanation section of the learning cycle is handled effectively, then these differences can be analyzed, compared, and made sense of, allowing for students to understand the concepts more deeply.

From a learning perspective, providing an initial and shared experience makes good sense. Instead of making guesses about the experiences students bring to the classroom (and undoubtedly not knowing with certainty for each and every child), the teacher can start with an experience and build on it. No students are left behind because the teacher assumes too much about what they already know. From a learning and teaching of science perspective, the direct experience is an essential feature of quality science teaching.

Although we have high regard for an inductive approach to science teaching, particularly in diverse classrooms, we must acknowledge that this approach represents particular challenges for some of the students in our classrooms. It is important to remember that asking questions about ideas, experiences, other learners, and the authority of the teacher is a very unfamiliar practice for many students. Some of your students may come from cultures and homes in which young people are not expected to ask questions, and for these students asking questions may seem disrespectful. At the very least, asking questions may be a little used skill for these learners. Because some of your students may not come to you equipped or comfortable with questioning skills, you need to allow for the development of such skills, perhaps by modeling such questions, explicating the need for questions in the doing and learning of science, allowing for lots of opportunities for students to question each other, and rewarding these efforts. Through such actions, teachers can help construct a culture of questioning in their classroom, something all students can benefit from.

Experiential Education

Using experiences within science teaching seems to be a way to move away from the drudgery of traditional instruction. We might begin to believe that direct experience for students is what will make all the difference in their science learning. However, experiences alone are not enough—even sitting at a desk listening to a lecture is an experience for the student. Maybe it's not an exciting experience and maybe it's not an experience that has any positive and lasting learning benefits, but it qualifies as experience.

How can a teacher decide which science experiences are likely to help students learn? How can a teacher determine if a particular activity is nothing more than fun and may not contribute to improved scientific understanding? The legendary educator John Dewey wrote about this issue in his 1938 book *Experience and Education*. Dewey tried to differentiate "traditional" teaching from "progressive" teaching—two opposing forces that are a source of tension in education even today.

What Qualifies as an Educational Experience

The quality of any experience has two aspects. There is an immediate aspect of agreeableness or disagreeableness, and there is its influence upon later experiences. The first is obvious to judge. The effect of an experience is not borne on its face. It sets a problem to the educator. It is [the teacher's] business to arrange for the kind of experiences which, while they do not repel the student, but rather engage are, nevertheless, more than immediately enjoyable since they promote having desirable future experiences … . Wholly independent of desire or intent, every experience lives on in further experiences. Hence the central problem of an education based upon experience is to select the kind of present experiences that live fruitfully and creatively in subsequent experiences. (Dewey, 1938, pp. 27–28)

Apparently Dewey saw a problem with simply rejecting traditional approaches to teaching. For all the right reasons, a teacher might vow never to teach in a traditional way. But Dewey noted that a teacher is not left with a new strategy or philosophy simply by discarding the one that exists. The statement "I'll *never* do that in *my* classroom" doesn't tell us (or the person who says it) what he or she will do instead. When it comes to experience, Dewey gave some guidance: the experience must be enjoyable and it must have connections with subsequent experiences. In other words, a science experience that is simply fun will not be sufficient. What is also required is that this experience continues to inform the students' understandings long after the materials are put away.

FROM PHILOSOPHY TO PRACTICE

As we discussed in the previous chapter, leaving students to freely explore (the discovery approach) is limited in its ability to support student learning. In the 1960s there was an explosion of science curriculum reform as the National Science Foundation provided considerable financial support for the development of new science education materials, many of which were inductive in nature. One of the elementary grade programs was called the Science Curriculum Improvement Study or SCIS. This is where the **learning cycle** was born.

The original learning cycle, designed by Robert Karplus, included three phases. In the first phase, students worked individually or in small groups with scientific materials as they pursued a problem or question. In the second phase, the entire class gathered together and the teacher orchestrated a discussion where the scientific concept was introduced. The third phase was an opportunity for students to apply this newly formed concept to different materials.

The learning cycle is still alive today, and we see it as an incredibly powerful model for teaching science. Over the years there have been modifications to the Karplus model, but it remains essentially the same. Direct experiences occur before concepts are taught. Presenting the concepts to the students is not enough—the students need to apply their ideas to see how well the concept and their understandings of it will transfer somewhere else.

If you do a search of "learning cycle" on the Internet, you will find a great deal of information, some of which will differ from what we present here. This is because the learning cycle has become almost a living thing, and in different environments, such as the business world, it seems to change its form. Regardless of the number of phases in a learning cycle (and we use a version consisting of five phases; see Figure 4.2), the model is fundamentally the same. Later in this chapter we will describe the learning cycle's appropriateness for teaching science to a wide variety of students. When Karplus was creating science materials, there was not much emphasis on multicultural education. We feel that the modifications to his original design can make this an effective approach for all students. Let's begin with the basic ideas and components of the learning cycle.

Phase One: Engage

The first phase of a learning cycle lesson is similar to many other approaches to teaching: you need to obtain the students' attention and orient their thinking toward the science they are about to study. If the students are working on another subject, the Engage phase helps them shift mental gears. If the students are returning to their room from another activity or entering from another classroom, then the Engage phase allows them to settle into science mode.

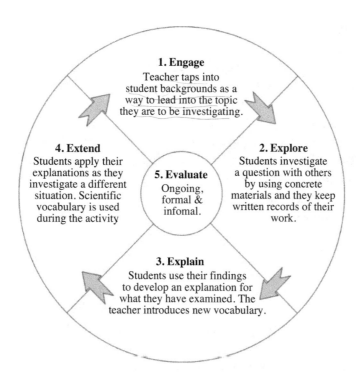

FIGURE 4.2: The learning cycle teaching model.

There is not a single technique that you will always use during the Engage phase. For a lesson that is a follow-up to previous lessons, the Engage phase could be in the form of a quick review of the activities the class has been doing. This review could be entirely under the teacher's control, or it could begin by inviting students to recall the previous lessons.

If this is the first lesson in a much larger science unit, the Engage phase is the teacher's initial opportunity for the students to begin making connections between their backgrounds and the science concept. In this case, the teacher could pose a question to students that taps into pre-existing knowledge. For example imagine a unit that deals with basic chemistry, a very common part of many elementary school science programs. On the first day of this unit, the teacher would want to connect his or her students' experiences with the topic. A way to begin would be to ask, "I want to start today by having you think about a time where you've watched somebody mixing a powder with a liquid. This could be when someone is cooking or doing laundry or lots of other things. But I want you to think about what happens when someone adds a powder to a liquid."

It is difficult to imagine the child who cannot think of an example of this phenomenon: adding sugar to a drink, pouring powdered detergent into a washing machine, mixing flour with water, putting fertilizer into a watering can, and so on. The intent of asking this question during the Engage phase is to honor students' personal experiences and use those as building blocks for the lesson. When teachers begin science lessons this way, there is a greater likelihood the students will be motivated to participate in the activity and that the subject matter will seem relevant and worthwhile.

Even if we tried, we could not list all the potential ways of doing the Engage phase. But here's a sample: displaying an object that is intriguing or playing a video clip that captures the students' attention. While students engage with these, invite them to talk about what they are noticing and what they are thinking. As is particularly appropriate in diverse classrooms, record these ideas using simple, straightforward words on a word board, an overhead projector, or some other central place in the room where these words and ideas can remain. The intention is for the teacher to allow the students to identify something that they already know or have wondered about and use that as a springboard into the science activity, as well as an entry point into the language of science.

Phase Two: Explore

The purpose of the Explore phase is to provide the students with firsthand experience with the science concept they are to learn. There is a delicate balancing act to this phase. We don't want to leave this open ended as if it's discovery learning, because we intend for this activity to connect with subsequent experiences. On the other hand, we don't want to make this so highly structured that it turns into a cookbook activity.

It is important to note that in the Explore phase, just like in the more structured types of inquiry, the teacher supplies some structure to the activity as a way to support the students' investigations. This includes pointing out the safety considerations, even if they are as simple as cleaning spills or not looking at the sun with the magnifier. In addition to such practical issues, the teacher should provide a challenge, problem, or question to guide the students as they explore.

Another indispensable way to provide structure is to give the students a record sheet. In working with teachers who use the learning cycle, we have seen that students' memories of the Explore activity are much weaker when they don't use the record sheet. This shouldn't be a fill-in-the-blank worksheet; it ought to provide places for students to record their observations. This could be as simple as "write three things you discovered when you mixed the powder in the water." The record sheet could also include a blank chart or table where the students enter their measurements. There could be a place where they are to draw what they observe. A record sheet is also a good way to encourage students to make inferences and predictions. For English language learners, it is helpful to have these worksheets written in English and their primary language. It is very common in school systems that serve a substantial population of non-native English speakers to have translation services provided for students in need. Also current curriculum materials often include handouts written in Spanish. Finding access to such services or establishing other community-based resources (e.g., bilingual members of the community willing to contribute time) at the outset of the school year will be an essential component of structuring successful experiences for English language learners. Having these students record their thoughts in their primary language and in English is an important way to help them learn the content. For students struggling with linguistic or cognitive barriers, drawing can be an excellent way of activating what they know and what they are thinking about the material. The point here is to create record sheets as vehicles for sponsoring student thinking in whatever way seems appropriate.

POINT–COUNTERPOINT

From the perspective of everyone who seems to have an official position about science teaching, inquiry seems to be the universal key. However, there are some who feel the definition of *inquiry* is so broad and vague that it's hard to determine what would not count as inquiry. In contrast, there are others who've witnessed teachers implementing inquiry and, despite the challenges, see the value to the students in those classrooms. So, in response to the question, "How useful is inquiry in real classrooms?" we have slightly divergent perspectives.

Integrating Inquiry into the Classroom: A Suspicious Construct?

Eleanor Abrams
Science Teacher Educator at the University of New Hampshire

Reform efforts tout inquiry as the way to learn science (American Association for the Advancement of Science, 1990; National Research Council, 1996). In fact, I feel that inquiry in the classroom has achieved such popularity with the so-called experts in science education that it makes other instructional techniques seem inadequate in helping students learn or, worse, makes it appear that all other approaches stifle students desire to learn science. However, as a science teacher educator, I have some nagging questions that need to be answered before I jump wholesale into advocating the use of inquiry into the teaching practices of science teachers: What does inquiry look like in the classroom? What teaching goals are enhanced by using inquiry as a pedagogical approach? What do you need to know as a teacher to integrate inquiry in the class effectively? How do you prepare and support your students to succeed during an inquiry-based project? These questions are left unanswered or, worse, the answers we get are often so vague they aren't of any help.

Take for instance the strategic way the National Research Council (2000) defined inquiry in the classroom. They outlined inquiry in terms of five components: who asks the questions, who designs the experiment, who manipulates the data, who draws the conclusions, and who communicates the results. These components are placed on a continuum from teacher directed on one end to guided inquiry in the middle to student generated on the other end. We can score our curriculum, our approach, or our assessment on how inquiry oriented it is. This approach reflects the complexity of trying to integrate inquiry into the classroom. However, it leaves me unsatisfied on the definition of inquiry because it seems that every piece of curriculum, every instructional methodology, and every assessment can be fit into this understanding of inquiry. If inquiry is everything, how can the notion of inquiry guide my science teachers' practice in the science classroom as they struggle to meet the very concrete and defined state science standards and high-stakes testing?

I think one of the pieces of information science teachers will need to integrate more inquiry-oriented approaches into the classroom is a "thick" description of when is it and where is it appropriate to use inquiry. When are other methodologies more effective in creating equitable, caring learning environments where all students can be successful learning science? The reform documents imply that teachers should strive to create an inquiry-based, student-oriented classroom for all learning situations. I am still waiting for the research evidence that says that inquiry is the only way to help students learn effectively.

There are other obstacles besides not being able to find a definition of inquiry or knowing when to use the technique to stimulate student learning. Teacher preparation courses, including my own, teach about inquiry, but most students have not experienced open-ended inquiry in their own science preparation. So I watch my teachers struggle to do what they have not seen modeled. Working in the dark, they take a leap of faith and integrate inquiry into the classroom knowing they are responsible for the very real demands of teaching content to their students. Some of my teachers have successful experiences and feel successful in the ways they do use inquiry in the classrooms. Others do not feel their students learn enough for the amount of time it takes, so they shun inquiry.

One of the unknown factors that determines the success of my teachers is whether their students are prepared and supported during the inquiry. Not a lot is written about what kinds of knowledge or skills students need and how that varies the kind of inquiry that is used. The little that is written indicates students resist the change from more teacher-directed approaches to more student-oriented methodologies because they don't know the rules for success. I think it takes a determined teacher to give the necessary information and skills to the students at the right time before and during the inquiry process, support the students' emotional needs as they experience inquiry for the first time, and articulate the benefits of this approach to parents, administrators, and other colleagues—while feeling uncertain that inquiry will yield the benefits in learning foretold by the science education experts like myself.

I argue that the science education community needs to become more sophisticated in its ability to define inquiry in the classroom, discern the appropriateness of the use of inquiry as compared to other pedagogies, and gather rich examples of how inquiry works so that science teachers are able to adjust their curriculum, support their students, and articulate why they are using inquiry in the classroom. Otherwise, teachers can look upon this notion of "inquiry" only with suspicion.

Inquiry-Based Science Can Be Done

Margaret R. Blanchard
Assistant Professor at North Carolina State University and Former Middle School Science Teacher

I can tell you this from my time in classrooms and working with teachers: not only can inquiry-based science be done but it is being done, beautifully so in some cases.

My dissertation was a study of secondary science teachers who were translating a field-based research experience to their classrooms. The teachers had a level of confidence born of the support they garnered through a research experience with other teachers. Each attended a five-week-long marine ecology program, in which they not only conducted science inquiry but also developed a lesson that modeled their learning into pedagogical practice.

What were these teachers' obstacles when they returned to the classroom? They varied. For one teacher, Nate, it was letting go of the notion he had to cover his wave lesson as "efficiently" as he did last year. For another, Tony, it was learning how to build a water rocket launcher and figuring out how to assemble enough supplies for four class periods of students. For yet another teacher, Trina, it was expanding her notion of what special education students were capable of (a lot more than she had expected!). Letting go of some of her class "control" was a struggle for Seville, a teacher whose middle school students conducted an inquiry lesson on light and color.

What set these teachers apart was not some mystical abilities or extreme levels of competence in subject matter. These teachers took the time and had the motivation to try something new in their classrooms. In some cases it was an extreme makeover, in others it was a new twist on an old version of the same content lesson. What happened when they did it? The students seemed more enthused with the lessons, the teachers found their time to be less rote and more stimulating, and the classroom power shifted. I remember getting goose bumps while listening to the quality of student questions asked following a presentation on a water rocket investigation. In my research, I found that inquiry fostered higher-order questions on the part of students and tended to shift the role of the teacher to a supportive role in the learning rather than as the dispenser of knowledge. For students to excel on high-stakes testing, their answering questions that ask them to think and solve problems and having experiences in inquiry-based learning seem essential.

Inquiry-based science teaching is not only possible but also essential if we are to teach about the processes of science instead of the usual products that are discussed in textbooks and traditional science lessons. I think that the National Research Council's definitions of inquiry make the process clearer than any other document I have ever seen. They give us outwardly observable behaviors that are a litmus test for who owns the learning in that classroom.

Does inquiry-based teaching require support? Yes. Time is essential: time for learning and planning on the part of teachers, and time for thinking, designing, and experimenting on the part of students. Institutional support is needed for this and for materials. But can it be done? Should it be done? Absolutely!

Phase Three: Explain

This third phase of the learning cycle has two parts. One part is the students communicating with each other about what they did, and the other part is the teacher identifying the concept that they have been studying. Both of these constitute the Explain phase. The students explain what they have found, and the teacher explains what to call it.

Because the students work in small groups or even by themselves, the Explain phase is important so they can learn what others have found. Inductive teaching is happening when the separate pieces of experience are pulled together into a unifying whole. In some instances it might make sense for the teacher to provide different materials to each group during the Explore phase. A plant unit that begins with the study of seeds might involve providing various fruits and vegetables to different groups of students. The Explain phase is the time in which students then compare and contrast materials. This initial aspect of the Explain phase is important for diverse learners, and the way in which groups are organized becomes very important. When your classroom includes English language learners, consider grouping pairs of them with able English speakers. Likewise organize students with cognitive limitations in heterogeneous groups. Such co-learning has been found to be a fundamentally useful manner in which to learn both science concepts and language.

The teacher's role during the Explain phase is to connect the students' learning to broader ideas. If the activity has provided enough structure, the students will have encountered the science concept at hand during the Explore activity. The teacher then attaches the label to the concept. What we are trying to avoid is thinking that the terminology is the same as the concept. The teacher isn't withholding anything by waiting to introduce the science term. The learning cycle allows the teacher to begin to convey the concept to students without burdening them with too much emphasis on correct vocabulary too early in the lesson. This is essential in everyone's learning of science, including English language learners and students with cognitive limitations. It's similar to a situation where you are familiar with someone or something but you don't know the right name for it. Although it's nice to know the name, the understanding of the person or object is not completely dependent on the name. And once you are familiar with the person or object, remembering the name is so much easier and more meaningful. In this way you should anticipate that the understanding and the label come together during the Explain phase.

In diverse classrooms, it is particularly important, at the end of the Explain phase, to move beyond simply discussing explanations or ideas as a class. Because some of the students may not have followed all the parts of the class discussion or reading assignments, perhaps because of simple language barriers or the cognitive complexity of the ideas, it is important to record the understandings the class has constructed and make them apparent and available to everybody. Multiple representations—charts, drawings, role playing, bilingual signs, and straightforward explanations in

English—all should be considered and employed to optimize students' science and language learning. Again, using a variety of ways to communicate the objects of science allows English language learners a better opportunity to understand the science and the methods to communicate it, and it allows students with cognitive limitations to better understand and remember what has been said.

Phase Four: Extend

This is the point where deductive teaching takes over. The students understand a scientific concept built during their recent activities, and they now even have a label for this concept. Their task during the Extend phase is to apply this understanding to a new situation. The parts from the Explore phase contributed to the whole from the Explain phase—that's inductive. Now in the Extend phase, students take this whole idea and test it against a new experience, which represents deductive thinking.

> **Moving From Induction to Deduction**
>
> While induction moves from fragmentary details (or particulars) to a connected view of a situation (universal), deduction begins with the latter and works back again to particulars, connecting them and binding them together. The inductive movement is toward discovery of a binding principle; the deductive toward its testing confirming, refuting, modifying it on the basis of its capacity to interpret isolated details into a unified experience. (Dewey, 1910/1991, pp. 81–82)

For whatever reason, teachers often feel tempted to push the Extend phase into new conceptual territory; that's not the right thing to do. Even though the teacher might be ready to move on to something new, the students are not yet there. During the Explain phase, the teacher points out the similarities across the Explore activity discoveries and uses scientific terms in describing the concepts. However, although a teacher will certainly have mastered the concept, the students will still need to invest time and experience to reach the same point.

During the Extend phase students take their new knowledge and apply it to another situation. Suppose that during the Explore phase of an activity, students mix water with salt, baking soda, and sand. During the Explain phase they would share and compare their results, while the teacher would introduce the term *solution* and use the students' own findings to reinforce the concept. Likewise during this phase the students would be expected to show that they could use the word *solution* correctly as they investigate new substances such as sugar, pepper, and powdered drink mix. The expectation is that this activity will reinforce the appropriate use of the vocabulary. What is of equal, or more, importance is the Extend activity should strengthen the students' grasp of the concept.

Phase Five: Evaluate

This last component of the learning cycle is a little bit different from the others because it can happen at several points within the learning cycle. Researchers have looked at the previous four phases and found that when the sequence of these phases is changed, the students learn less well than when the sequence of phases occurs as we've described (Abraham & Renner, 1986). In contrast, the Evaluate phase can legitimately occur at several places within the learning cycle.

For right now, let's treat the Evaluate phase as the last in the sequence of learning cycle phases. This is the place and time in which the teacher evaluates what the students have learned. It also can become a way to inform the students about how well they understand what's been studied. This should somehow be formalized. Evaluation might not be as formal as a written, multiple-choice test, but it shouldn't fall to the other extreme and be based on casual observations of the students as they work. Teachers must be attentive to the type of the evaluation they use in order to minimize the tension between getting at what students understand and optimizing their opportunities to express what they genuinely know. Think in terms of multiple modes of communication, and carefully consider drawing, acting out, graphing, and writing as well as using more standardized measures.

FOR REFLECTION AND DISCUSSION

How might the teacher's interactions with the students be different during the various phases of the learning cycle? What is the teacher listening for? What sorts of questions might the teacher ask? What is the goal of the teacher during the different phases?

As with all types of assessments, the Evaluate phase should closely align with the information students have been given and the activities in which they have been participating. Expecting students to respond to a written quiz can be appropriate as an Evaluate phase activity if it requires them to employ the same thinking processes, scientific vocabulary, and process skills they have recently been using. However, an Evaluate phase activity that is a dramatic cognitive departure is not appropriate within the learning cycle. Additional details regarding assessment can be found in Chapter 7. But for the sake of this discussion, let's position the Evaluate phase at the end of a chain of learning experiences.

APPROPRIATENESS FOR ALL STUDENTS

The learning cycle came into being around the time that discovery learning and inquiry were in vogue. Our sense of the limitations of those two approaches doesn't apply to the learning cycle. First, the learning cycle begins with an emphasis on the children and the act of guiding them to think scientifically, rather than having the scientist as the starting point. This approach avoids the danger of excluding students because of any unexamined assumptions about the types of kids who can and should do science.

Another benefit of the learning cycle is that in the Engage phase it makes a deliberate effort to connect with students' prior experiences. Also, in a classroom where the teacher is not completely aware of each child's family background and cultural traditions (which means almost every classroom), the Explore activity provides a shared experience. Then as the students talk about their science notions, they have something in common. No one is left out because the example being used is foreign to his or her experience. And although the students may have a different interpretation of the experience given what they bring into the classroom in terms of their background, the explain phase should be structured in such a way to make these differing interpretations explicit through comparison and discussion—providing another opportunity for students to refine and deepen their understandings.

A third aspect of the learning cycle that makes it suitable for use with all students is the manner in which learning takes place. Instead of the information coming directly from the teacher, the students are able to test their ideas as they converse with their classmates. Some students have a difficult time asking questions of teachers—this can be because their culture regards the questioning of adults as disrespectful. By creating a classroom culture where students are often working with their peers, questions are raised and debated without the danger of confronting authority. This sort of interpersonal communication is especially prominent during the Explore and Extend phases as the students work on activities.

The fourth aspect of the learning cycle that makes it suitable for diverse learners is inherent in the combination of inductive and deductive approaches. Encouraging inductive reasoning from the outset allows students to gain experiences with which to associate terminology, a definite plus in terms of how people learn. The deductive aspects provide another opportunity to consider these ideas and to further link the experiences to the scientific explanations—again a benefit in terms of enhancing the learning of all students regardless of their English fluency or cognitive abilities.

EXTENDING SCIENCE TO ALL LEARNERS

We continue examining the Principles for Effective Pedagogy proposed by the Center for Research on Education, Diversity, and Excellence (Warren & Rosebery, 2002) for their application to science teaching. The third of the five principles is "making meaning: connecting school to students' lives," and it continues to foster the notion that teachers can help all students be successful in science. The basis of this principle is that teachers' backgrounds are often very different from those of their students. As a result, students from diverse populations may have difficulty recognizing the connections between the content presented in school and the realities of their personal lives. For English language learners this challenge is compounded because of the lack of language in which students and teacher are mutually fluent. The teacher's challenge in making connections is to unearth, appreciate, and make effective use of the students' sources of knowledge.

Connecting the World of Students to School Science

In examining Table 4.1, which lists approaches to building connections between school and home, you may recognize the overlap with the learning cycle. The first teacher action in the list parallels the Engage phase by starting with ideas already familiar to the students. As we've expressed before, this principle, as well as the others, seems appropriate for all students, and not just those who are developing English fluency. Nevertheless, it is necessary to provide students with opportunities to make these connections. Leaving such learning opportunities to chance gives no assurance that they will happen. In addition, such strategies give substance to the claims a teacher might make about being genuinely interested in his or her students. It's one thing to say "I care," but it is much more convincing when the actions of the teacher reinforce such statements.

TABLE 4.1: Connecting School to Students' Lives and the Appropriate Teacher Actions

THE TEACHER:

uses what students already know from home, family, and neighborhood to shape the activity;

uses local community norms and knowledge sources to design meaningful learning activities;

guides students to relate and use what they've learned in their home community; and

varies activities in structure to accommodate student preferred forms of participation: collective versus individual and cooperative versus competitive.

Note: Adapted from Dalton (1998).

The other components of this, the third principle for effective pedagogy, provide further justification for the learning cycle as an appropriate mode of teaching science to students who are English language learners. In particular, the Extend phase seems to be a rich opportunity to take knowledge acquired in the classroom and find ways to apply it to the neighborhood. In fact, all of the strategies listed for this third principle seem to mesh with the use of the learning cycle. The acts of finding ways to make meaningful links between academic science and the students' communities and modifying activities to allow for cooperative or individualized activities are well within the bounds of the appropriate use of the learning cycle. Such a synthesis would contribute to increases in the science learning of English language learners, with the potential for reducing the achievement gaps.

KEY FEATURES OF THE LEARNING CYCLE

The learning cycle is an approach to teaching that fosters strong student understanding of science concepts. In a science unit, the second learning cycle should build on its predecessor. For a science unit that consists of multiple learning cycles, we can envision a stack of learning cycles where the information in one provides the foundation for the next. This method of curriculum design has been called a **spiral curriculum**, which describes the path that the teacher and students take as they move upward through a growing stack of learning cycles. The expectation is that information is not simply learned once and left behind. Instead, as students move through the spiral, they circle back to previously learned information and skills, but in more complex and meaningful ways. This is a very important way of crafting the science experiences to enhance learning in classrooms with diverse students.

Another significant feature of a learning cycle is that only one concept is addressed within all five phases. Too often teachers feel they need to move forward and begin to present new concepts partway through. This is counterproductive. The desire to push ahead and move on is based in an old notion of content coverage. In too many situations teachers have been told that they have to cover an overly ambitious number of pages in the science book by a deadline. They have been told that if they don't keep up with all the other teachers, then the next year their students will be behind all the others.

The consequences of this folly are numerous. Students can become discouraged because they don't have enough time to really understand the material or gain the opportunity to express what they know. Teachers become anxious about keeping a good pace, and they transmit this as impatience toward their class. When this is the case, students don't have the chance to truly comprehend the important science concepts because the amount of time needed to struggle with and resolve ideas is never allowed.

We have made several interesting discoveries after making international comparisons of science teaching and learning. One that was widely reported in the media was that many countries have higher science achievement than the United States. Let's imagine for a moment that the comparisons weren't between the best students in other countries and typical U.S. students. What do you predict the textbooks of those other countries would look like? A logical way to think is that better test scores would mean knowing more material, which would be reflected in larger and more detailed textbooks. After all, wouldn't you expect that you'd gain more knowledge from a thick text compared to a slim volume?

If that was your prediction, then you would be wrong. The science textbooks in other countries where student achievement was higher were consistently and substantially shorter than the textbooks in the United States. In looking at the content of the textbooks, researchers discovered that in the United States, there is a great deal of information covered in every grade level but with only a surface treatment. In fact many of the same concepts were being taught from one year to the next. In contrast, the slimmer books used in other countries covered less material but in much greater detail. Furthermore, there was very little redundancy from one year to the next. One way to characterize this is to describe the science curriculum in the United States as "a mile wide and an inch deep." This situation has seemingly contributed to our students learning less science.

It is fortunate that the current science standards produced through reform movements in the United States have much greater depth while avoiding the tendency of too much breadth. For the most part it seems that individual states have followed that practice as they have adapted their own state curricula to these national reforms. In elementary and middle schools, there is less of a push to cover an entire textbook in one year. Students are expected to experience three to eight science units, and these units are not the same at each grade level.

This brings us back to the learning cycle. Because each learning cycle can address only one science concept and because it may take several class sessions to make one pass through the learning cycle, the amount of science material that can be covered in any one year is greatly reduced. With the learning cycle approach, the students spend more time developing deeper understandings and communicating what they are learning, even though they may not cover as much material. It is fortunate that the science education standards are consistent with this teaching approach. Rather than having the relentless drive to move from one chapter to the next in rapid succession, the science standards and the learning cycle have the potential to work in concert with each other.

VARIATIONS ON THE LEARNING CYCLE

The most widely recognized form of the learning cycle is typically associated with science educator Rodger Bybee (1997, 2002). Just as with the learning cycle we presented, Bybee's consists of five phases, each of which starts with the letter "E." When you hear or read about the "Five E" model, you should give appropriate acknowledgment to Bybee's promotion of this variety of the learning cycle. However, this shouldn't lead anyone to believe that all learning cycles have five phases.

The original learning cycle attributed to Robert Karplus was a three-phase model: exploration, concept introduction, and concept application (Atkin & Karplus, 1962). Even at its inception the exploration of materials by students clearly preceded the formal teaching of scientific concepts, and this in turn was followed by subsequent activity in which students applied these concepts to new situations. Just as with Bybee's learning cycle, the Karplus version served as the guiding instructional approach for a science program called the Science Curriculum Improvement Study (SCIS) (Karplus & Thier, 1967), which is still commercially available.

Another version of the learning cycle consists of four phases and is best illustrated with the Insights curriculum developed by the Education Development Center (EDC). As described within its materials, the EDC learning cycle proceeds from getting started, exploring and discovering, processing for meaning, and extending the learning experience

(EDC, 2003). The distinction from the Karplus model is the addition of a preexploring phase, and the difference from the Bybee model is the evaluation phase. However, the EDC authors don't ignore evaluation: their preference was to infuse it across all phases of the learning cycle and not force it into a separate phase.

Another variation of the learning cycle has been proposed. Eisenkraft (2003) inserted an Elicit phase before Bybee's Engage phase and appended an Extend phase after Bybee's Evaluate phase. The creator of the 7E model freely acknowledged that his model is built on the legacy of its predecessors. And he pushed toward seven phases in response to findings from cognitive research that stress the value of making students' initial ideas explicit (i.e., with the Elicit phase) and the necessity of having students apply their understandings to new scenarios (hence the Extend phase). It is admirable that Eisenkraft uses his learning cycle within the *Active Physics* curriculum he has created. A comparison of these models appears in Table 4.2.

SPECIAL NEEDS POPULATIONS AND THE LEARNING CYCLE

To the best of our knowledge, there has yet to be an educational research study that specifically examines the effectiveness of the learning cycle for special needs populations. Although there's no evidence the learning cycle would not work, there isn't a study we can point to that says it is better than other approaches. But there is no cause for despair, because there is considerable evidence that the underlying philosophy of the learning cycle and the manner in which this model is implemented in the classroom does correlate with demonstrably successful methods of teaching science to students with a range of cognitive abilities.

TABLE 4.2: Despite Variation in the Number of Phases, All Learning Cycles Have a Shared Structure

KARPLUS AND THIER (1967)	EDUCATION DEVELOPMENT CENTER (2003)	BYBEE (1997)	SETTLAGE AND SOUTHERLAND (2012)	EISENKRAFT (2003)
Exploration	Getting started	Engage	Engage	Elicit
Concept introduction	Exploring and discovering	Explore	Explore	Engage
Concept application	Processing for meaning	Explain	Explain	Explore
	Extending the learning experience	Elaborate	Extend	Explain
		Evaluate	Evaluate	Elaborate
				Evaluate
				Extend

The research that has been conducted investigating special needs students in the context of science has been the specialty of Margo Mastropieri and Thomas Scruggs. In their extensive review of research of the science educational experiences of special needs populations, they summarized their field with these encouraging comments:

> Accommodation of individuals with special needs in science classrooms does not necessarily conflict with current thinking about science education and may, in fact, be very compatible with several aspects of such thinking. The aspects including emphasizing general principles and overall themes over separate facts and vocabulary, deemphasizing text; providing more focus on less content; including concrete examples and hands-on activities; [and] promoting cooperative group solutions to scientific problems. (Mastropieri & Scruggs, 1992, pp. 404–405)

Although these recommendations are not based on specific investigations of the learning cycle, we recognize key features of this instructional model that apply to special needs students. For example the emphasis of general principles coincides with the "one concept" rule. The preferred focus on more in-depth exposure to fewer topics is consistent with the admonition to provide sufficient time to move through all of the phases. Concrete experiences and hands-on lessons are at the core of the Explore and Extend phases. Finally, the recommendation for having students work in groups is a defining feature of the learning cycle. On a point-by-point basis there are clear indications that the learning cycle is an appropriate teaching model for helping special needs students to become successful at science.

CONNECTING THE LEARNING CYCLE TO THE THREE SCIENCE TEACHING APPROACHES

In the previous chapter we presented you with three approaches to science teaching, whereas in this chapter we introduced you to a specific instructional model. A reasonable question one might ask is where are the connections between the approaches and this model. The confusion is understandable because, within discussions and writings about science teaching, people use the same term to signify very different philosophies, and people might be relying on different terms when they are in fact speaking of the same idea. Not everyone is aware of this mixture of meanings, and we are grateful to Bill Holliday (2001) for bringing this situation to the attention of a wider audience.

In an effort to help clarify what is meant by *inquiry* and how it contrasts with the discovery approach, Holliday recommended reading the *National Science Education Standards* and the supplement the National Resource Council published that concentrates on inquiry (National Research Council, 1996, 2000). But, as Holliday (2001) explained, "neither book presents a clear picture of the meaning of inquiry" (p. 55). He went on to speculate, "Perhaps this is because the term is increasingly fraught with different meanings" (p. 55). We see this as an open invitation to make the relationship between the three teaching approaches and the instructional model clearer.

We regard the three approaches as general schemes for presenting science to students but without necessarily specifying how this might be accomplished. In contrast, the learning cycle is much more specific about the delivery of instruction. Perhaps a sporting analogy will help make this clearer. In preparation for a basketball contest, a coach might decide to approach the next opponent with a great deal of intensity in an effort to establish a fast-paced game. Although that's a logical approach, it does not dictate how the players are to perform (beyond being intense). The strategies to actualize the approach might be to use a full-court press on defense and run a fast-break offense after every basket by the opponent and defensive rebound. The approach is rather general, but the actual strategy is quite specific. This is how we imagine the relationship between inquiry and the other two science teaching approaches and the learning cycle as an instructional model: the former describes the ambition whereas the latter identifies the ways it will be put into action.

We admit that it seems difficult to connect the discovery approach with the learning cycle model: discovery is characterized by its lack of structure and very little teacher directedness. However, both the inquiry and conceptual change approaches lend themselves to the learning cycle. Initially, the instruction builds on an inductive plan where experiences are used to build toward a significant concept. Subsequently, this concept is deductively applied to new situations. The difference between these applications is that the inquiry approach emphasizes the Five Essential Features while the conceptual change approach concentrates upon uncovering students' preexisting notions and then deliberately displacing those ideas that are not aligned with scientifically acceptable explanations. The learning cycle does not dictate which approach (i.e., inquiry vs. conceptual change) is used. In fact, the learning cycle could be appropriately used for a variety of inquiry approaches from structured to guided to open (Colburn, 2000).

A CAUTIONARY NOTE

When we've had this discussion with practicing teachers, when we've discussed the need to acknowledge, build upon, and address the very different understandings that students bring into the classroom, we've had teachers occasionally throw up their hands and say "that's too hard," "that's impossible," or "that makes teaching too complicated." Indeed we've both experienced these feelings in our own teaching. But we must acknowledge that actually helping students learn science in deep, useful ways is complicated. If learning depends on what the learner already knows (and it does), then teaching becomes a very difficult task. It is hard. It does take time. And, as we will discuss in Chapter 14, despite the enthusiasm we have for the learning cycle, we understand there is no single "right way" to help students learn. But by using some of the metacognitive strategies (which we will describe in Chapter 12 on interdisciplinary teaching) and by paying close attention to students' discussions during their explorations of phenomena, teachers can gain access to some of their students' thinking—then use that knowledge to better shape their instruction. One of the things we find so useful about the learning cycle is that it provides opportunities for teacher listening—as we understand that care paid to student ideas is an essential component of the teaching of science.

CHAPTER SUMMARY

- Teaching involves addressing the major ideas and specific examples of those ideas. When preparing for science instruction, the teacher must recognize the need to include both details and generalizations and decide the most effective sequence for their presentation.

- A deductive approach to learning begins with the major idea followed by the specific examples. The specific details are used to support the general ideas because the generalities are followed by the examples. In contrast, an inductive approach starts with the students working with particular examples that eventually lead to the major idea.

- By beginning a science unit with direct experiences, teachers provide all learners with a foundation for supporting their developing ideas. Shared experience allows students to construct more systematic, detailed, and useful understandings of concepts than is often possible through instruction that emphasizes vocabulary.

- The learning cycle combines inductive and deductive approaches to science teaching. Each trip through the learning cycle focuses on a single yet important science concept with a mix of direct experiences and whole class discussions and explanations of the scientific phenomena in question.

- Despite variations in the number of phases and the labels given to the components, all learning cycles share the same overall structure. The five-phase model used in this chapter reinforces the inductive and deductive components of effective application of this model.

- An inductive approach to science teaching supplies every student in the class with the same foundational experiences. Because they have shared experiences, students with divergent backgrounds and experiences have less of a disadvantage when the teacher begins describing the science concept.

- Although we see the learning cycle as an almost ideal approach for teaching science to diverse populations, it is important to recognize that even this approach has its limitations. Teachers need to recognize that questioning ideas or teachers comes more naturally to some students than to others. Thus it is important to slowly construct the culture in the classroom where students learn to question and teachers support such questioning.

- It takes time for students to learn important science concepts! As a consequence, the learning cycle is intended to concentrate on the same concept through all phases. It is inappropriate to introduce new concepts partway through the learning cycle in an attempt to cover more standards.

KEY TERMS

Deductive approach: a common approach to instruction in which the lesson begins with a general idea or concept and progresses to more specific instances of the idea or concept.

Inductive approach: an approach to science instruction in which the lesson begins with presentations of part of an idea or concept, typically through examples or activities but without explaining how the parts are related, and progresses to the teacher guiding students to understand how the examples and activities are related to a more general idea or concept.

Learning cycle: an instruction model that originally consisted of three phases: (1) students working with scientific materials to pursue a problem or question, (2) a class discussion in which scientific concept is introduced, and (3) an opportunity for students to apply this newly formed concept to different materials.

Spiral curriculum: a method of curriculum design that consists of multiple learning cycles, where the information addressed in one learning cycle provides the foundation for the next. The spiral describes the path that the teacher and students take as they move upward through a series of learning cycles.

SUGGESTED READINGS

Cavallo, A. M. L. (2001). Convection connections. *Science and Children, 38*(5), 20–25.

Within this article the author provides two exemplary learning cycle sequences. In the first sequence, students learn about convection currents in air, whereas the second sequence builds understandings of water convection currents. This demonstrates how one learning cycle can contribute to the learning that will occur in the next cycle.

McCarthy, D. (2005). Newton's first law: A learning cycle approach. *Science Scope, 28*(5), 46–49.

Newton's First Law of Motion is about inertia: objects at rest tend to stay at rest, and once they are moving, the object continues to move in the same direction unless another force interferes. All of this is embedded within this Louisiana teacher's use of the learning cycle. This author's version of the learning cycle consists of four phases: elicitation, exploration, invention, and application.

REFERENCES

Abraham, M. R., & Renner, J. W. (1986). The sequence of learning cycle activities in high school chemistry. *Journal of Research in Science Teaching, 23,* 121–143.

American Association for the Advancement of Science (AAAS). (1990). *Science for all Americans*. New York: Oxford University Press.

Atkin, J. M. & Karplus, R. (1962). Discovery or invention? *The Science Teacher, 29*(5), 45–51.

Bybee, R. W. (1997). *Achieving scientific literacy: From purposes to practices*. Portsmouth, NH: Heinemann.

Bybee, R. (2002). *Learning science and the science of learning*. Arlington, VA: National Science Teachers Association Press.

Colburn, A. (2000). An inquiry primer. *Science Scope, 23*(6), 42–44.

Dalton, S. S. (1998). *Pedagogy matters: Standards for effective teaching practice*. Washington, DC: Center for Applied Linguistics.

Dewey, J. (1991). *How we think*. New York: Prometheus. (Original work published 1910.)

Education Development Center (EDC). (2003). *Insights: An elementary hands-on inquiry science curriculum*. Newton, MA: Author.

Eisenkraft, A. (2003). Expanding the 5E Model: A proposed 7E model emphasizes "transfer of learning" and the importance of eliciting prior understanding. *The Science Teacher, 70*(6), 56–59.

Holliday, W. G. (2001). Critically considering inquiry teaching. *Science Scope, 24*(7), 54–57.

Karplus, R., & Thier, H. (1967). *A new look at elementary school science*. Chicago: Rand-McNally.

Mastropieri, M. A., & Scruggs, T. E. (1992). Science for students with disabilities. *Review of Educational Research, 62*, 377–411.

National Research Council (NRC). (1996). *National Science Education Standards*. Washington, DC: National Academy Press.

National Research Council (NRC). (2000). *Inquiry and the National Science Education Standards: A guide for teaching and learning*. Washington, DC: National Academy Press.

Warren, B., & Rosebery, A. S. (2002). *Teaching science to at-risk students: Teacher research communities as a context for professional development and school reform*. Santa Cruz, CA: Center for Research on Education, Diversity, and Excellence.

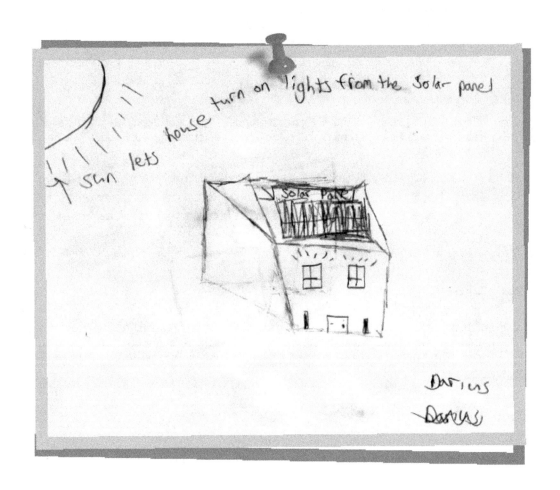

READING 5

LET'S EXPERIMENT!

By Ann Finkelstein

> Youth is wholly experimental.
> —Robert Louis Stevenson

Care taken in defining a scientific question and planning controls should result in an experiment that is easy to do and data that are simple to analyze. Doing the experiment involves keeping a laboratory notebook, collecting the data, and coping with experimental error.

WRITE EVERYTHING DOWN.

The conventional wisdom about keeping a student laboratory notebook seems to revolve around two main tenets. (1) Each new experiment should begin with the statement "The purpose of this experiment is to demonstrate ..." (2) Neatness is paramount. I suggest a different tactic. Writing the purpose of the experiment implies that the outcome is known in advance. If experiments are designed to test a hypothesis, then the conclusions will result from a comparison of experimental samples to their controls. Thus, a scientific question is not answered until the experiment is complete. An appropriate title for experiments can be the question that the experiment is designed to answer.

How important is neatness? The student must be able to decipher his notes at a later date. Other people should also be able to read the notebook. Laboratory notebooks, however, are in the thick of things. Reagents are spilled on them, they may be carried to snow-covered hills, and data are recorded while the experiment is in progress. The final presentation of the data should be neat, orderly, and beautiful, but as long as the notebook is a complete and legible record, its style is unimportant.

A good notebook gives the experimenter control over the experiment, and helps ensure reasonable results. Experimenters should plan their experiments in advance, and write the plans in the notebook. Knowing what you are going to do before you do it often eliminates unwelcome surprises in the laboratory. Valuable skills are developed in devising experimental protocols and learning how to follow them. Similarly, the ability to keep an accurate record of events has applications beyond the completion of an experiment.

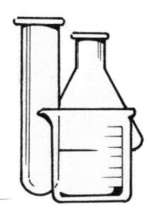

Can children keep a laboratory notebook? Students who are learning to write will need help, and older children need

Ann Finkelstein, "Let's Experiment!," *Science is Golden: A Problem-Solving Approach to Doing Science with Children*, pp. 57-65. Copyright © 2002 by Michigan State University Press. Reprinted with permission.

direction as well. The maturity level of the students dictates how much assistance they need. Children often like to keep track of information, so recording data comes naturally to them. For young children, compiling a laboratory notebook can be a group exercise, and brainstorming can be used to involve the children.

What should go into a laboratory notebook? Just about everything. Students should be encouraged to write everything down, and never trust their memories. A complete laboratory notebook helps the scientist do the experiment without mistakes, facilitates the analysis of the data, and helps with the presentation of the results. The following list should help students construct and maintain their notebooks.

1. *The date*. Record the date when the experiment was done.
2. *A title*. The question the experiment is designed to answer can serve as a title.
3. *Answers to the list of questions about experimental planning* from Chapter 3. If brainstorming was used to plan the experiment, a condensation of the relevant ideas can go in the laboratory notebook. Most importantly this should include:

    - Information obtained from library research. See the list of facts on friction in Chapter 3.

    - The model. The student should state how she thinks the system works.

    - Considerations about experimental consistency. For the sledding experiment, there should be notations regarding timing the runs, having the same person ride all the sleds, and testing all the sleds three times.

4. *Controls*. List the negative and positive controls that will be included in the experiment. Write down the assumptions that the negative controls are designed to test. What essential components are omitted from the negative controls? Which experimental conditions are supposed to "work"?
5. *Materials and methods*. What is needed to do the experiment? Another scientist wishing to duplicate this experiment should be able to find all of the necessary information in the materials and methods section of the notebook. The list of materials for the sledding experiment would include descriptions and sketches of the sleds, and a notation of any other equipment (stopwatches and calculators) used in the experiment. The methods section should include a detailed report of how the experiment was done. A laboratory notebook for the sledding experiment would include a description of how the sled run was prepared, the name of the rider, the number of pushes the rider used, the length of the hill, how the timing was done, etc.

6. *The data.* Depending on the experiment, the data may be:

    - A table of numbers. (In the sledding experiment, the data would be the times required for each sled to complete the run.[1])
    - A qualitative analysis. (In the dirty sock and laundry detergent experiment, brand A might be rated ++ for cleaning power and brand B rated +++.)
    - A detailed description. (A laboratory notebook for the stain solubility experiment could include a description of the final appearance of the coffee filters, a record of how far the spots moved, and whether the spots changed color as their components separated.)
    - Photographs or sketches. (A photograph showing the washed and unwashed socks would clearly demonstrate the results.)

7. *Notes about the experiment.* Any information that is relevant to the analysis of the data should be included. For example, students doing the sledding experiment should note that one sled could not be kept on course because it was so difficult to steer. Perhaps one laundry detergent turned the socks yellow, or ate holes in them. Any problems that occur while doing the experiment should be recorded in the notebook.
8. *The analysis of the data.* The laboratory notebook should include sample calculations, tables, and graphs. Data analysis is covered in detail in Chapter 6.
9. *The conclusions.* Was the question answered? What was learned?
10. *Notes on future experiments.* Did the experiment lead to more questions? How could they be answered?

Loose-leaf notebooks and notebooks with perforated pages should not be used as laboratory notebooks because pages may be lost. Expensive bound laboratory notebooks with graph paper and carbon copies are not necessary. A simple spiral pad works well as long as extra pages such as graphs, photographs, sketches or computer printouts are taped securely in place.

A laboratory notebook for the bean seedling experiment is provided as an example in Appendix 2.

BE CAREFUL. BE CONSISTENT.

The experiment should be done as carefully as possible. Small inconsistencies may result in big changes in the data. There is an element of craft in doing experiments and producing beautiful data that can generate as much personal satisfaction as the ability to hit a three point shot or to do the splits. Of course, inexperienced hands may make mistakes especially in the excitement that often accompanies children's experiments. Making a mistake is not the end of the world. Children should be encouraged to take the collection of data seriously, but almost any experiment can be repeated, and perhaps improved in the repetition.

Label all reactions, trials, and controls. Don't trust your memory to distinguish the sock washed in brand A from the one washed in brand B. In the controlled chaos that characterizes children's experiments, samples can be mixed up. When the smoke clears, it will be easier to analyze and double check labeled samples. Do not label the lids of containers, as lids can be removed and separated from the experiment. For samples that are difficult to label, try a creative approach. If several bean seeds are planted in one cup, record the location of each seed with a different colored mark on the rim of the cup. The pennies in the lemon juice experiment can be labeled on masking tape attached to the flip side. "Permanent" markers contain ink that is not soluble in water, so they are often useful in wet experiments.

Safety is always a concern when doing science experiments with children. An adult should handle hazardous chemicals such as strong acids and bases. All persons involved in experiments using potentially dangerous chemicals should wear eye protection, rubber gloves, and lab coats. Experiments that require flames, heat, or electricity should also be monitored closely by an adult. Broken glassware and other sharp objects are potential hazards. In most cases, common sense can ensure that experiments will be safe, educational, and fun.

Well, what about fun? Actually doing the experiment is often the most fun part of the scientific process. It's time to take chances, make mistakes, and get messy. What could be better? Much time and effort has been spent on the experiment since the children first asked the question. Finally learning the answer should be the moment they have all been waiting for.

Leave the anxiety at the door. Because the experiment has been planned and done as a group effort, the stress component should be reduced. The teacher or parent should not feel compelled to demonstrate a perfect experiment. The students should not worry if the data proves their model wrong. The point is to learn something, and learning *is* fun. An elementary school teacher once admitted to me that the chemistry class she had taken in high school had "provided enough stress to last a lifetime." I can remember experiencing a sinking feeling in my own laboratory courses when I realized that I had botched the experiment, and was not going to have enough time or reagents to repeat it. Mistakes happen. Schedule enough time and procure enough supplies so that the experiment can be repeated (and improved) if necessary. Be flexible. It is better to spend a little more time and end up with a successful learning experience, than to hurry on to another subject before the experiment has been done correctly. Creativity is stifled in a tense environment. Real scientists do not produce successful experiments every day. Why should we expect that of children?

Be consistent. Remember what the question is asking, and what the experiment is testing. No other factor should be varied between experimental trials. The sledding experiment was designed to compare the differences between sleds, so conditions were kept as similar as possible from run to run. The same rider made all the runs, the rider pushed off the same way each time, and the length of the run remained the same. The experimenters anticipated that the condition of the snow might change as the experiment progressed, so all the sleds were tested three times, and every sled was tested on packed snow.

The best experiments for children test one thing at a time. The experiment on bean growth is comparing different soils, therefore, all the plants should receive the same amount of water and light, they should be grown at the same temperature, and the pots should all hold the same amount of dirt. The penny brightening experiment is investigating how much time is required to remove tarnish from copper. To ensure consistency, all pennies should be about the same color to start with, and the juice drops should all come from the same lemon. Strive for similarity in all aspects of the experiment except the one that is the subject of the experimental test or control.

LIFE IS IMPERFECT, AND SO ARE YOUR DATA.

Very few experiments work perfectly, although people seem to expect perfection on the first try. Numbers are rarely "spot on," there are glitches, and there are differences between apparently identical trials. Why? There are three main reasons for imperfection. (1) An unknown variable changes between trials. For example, students with little sledding experience may not have anticipated that the hill would get faster with successive runs. (2) Biological variation is also a source of "error." No two living things are exactly alike, even identical twins. Two seemingly identical beans planted in identical conditions may germinate on different days. Biological variation is part of life. These little differences make the world interesting, and allow species to adapt and evolve. (3) Small human errors in making measurements or performing the experiment can affect the data. If the person timing the sled runs did not stop the stopwatch exactly when the sled crossed the finish line, that sled would appear to have a slower time. Glitches are okay; they are inevitable; we need to cope with them.

How do scientists cope with all these errors and differences between samples? They use repeated trials. Repeated trials can aid in understanding variations caused by unknown variables, biological variation, and human error.

An example of a discrepancy caused by an unknown variable can be found in a popular science fair experiment. Children seem to love experiments designed to determine the favorite food of a family pet. (Family pets probably love these experiments too.) An experiment was planned to test whether a guinea pig preferred carrots or cucumber slices. Every morning for a week, the experimenter put the two foods in opposite corners of the cage, and recorded which vegetable the guinea pig ate first. Let's pretend that

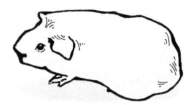

one morning the experimenter was unaware that her younger sister had already fed the guinea pig, and on this day the sated rodent was indifferent to the treats. If a single trial had been used, the experimenter might erroneously conclude that the guinea pig liked neither carrots nor cucumbers. As the experiment was repeated each day for a week, the anomalous day stands out as a "glitch." That trial still needs to be recorded and analyzed, but its effect on the conclusion is less pronounced.

Similarly, the best way to deal with biological variation is to increase the number of people, animals, or plants studied. Imagine an economic experiment designed to test if brand-name jeans last longer than generic jeans when worn by boys in the first grade. Special emphasis will be placed on the survival of the fabric in the knee region. The differences between boys will undoubtedly be greater than the differences between fabrics. Some boys are more active than others. The types of activities will also vary, some will be kneelers, some will be sitters, and some will be runners and/or fallers. One way to circumvent this problem is to test lots and lots of boys. If there are enough boys in each group, the differences between boys will "average out." In other words, on the average, each group will contain about the same number of active and not-so-active boys.

Human error is always a factor in experiments. Care and skill can minimize this source of variation, but it will always affect the data. Another experiment is designed to test if iron-on or sew-on patches last longer when used to mend little boy's jeans. The adhesive on the iron-on patches may be affected, for example, by small variations in the temperature of the iron. Even if the experimenters are taking care to be as consistent as possible, small variations in iron temperature may have noticeable effects. Similarly, the length

of the stitches used to attach the patch may affect the longevity of the sew-on variety. Experimenters sewing patches by hand will not be able to produce absolutely uniform stitches for every patch. The best results will be obtained by increasing the sample size. If enough patches are analyzed, small differences in the way the patches were attached to the jeans, will again "average out."

What about experiments that just do not "work"? These are not experiments that proved that the model was incorrect. These are experiments that produce inconclusive data or no data. What does the experimenter do now? *If at first you don't succeed, try, try again—but first think about what went wrong.* What can be changed to improve the experiment? I tried most of the experiments discussed in this book. Not all of them worked the first time, but I sometimes learned valuable information from the "failures." From the apple browning experiment in Chapter 7, I learned that tart, green apples do not turn brown even when they are exposed to air for a long time. When testing the stain solubility experiment in Chapter 4, I learned that the ink in ballpoint pens is not soluble in water. (Ballpoint pens are therefore useful for labeling the coffee filters.) The first time I did the bean seedling experiment (Chapter 4 and Appendix 2), only one bean grew, so I repeated the experiment with four times as many seeds. A failed experiment is not the end of the world. A failed experiment can provide an opportunity to learn. Children may be disappointed when their experiment does not yield analyzable data. Try brainstorming to come up with new approaches to the experiment. Positive controls are particularly useful for troubleshooting unsuccessful experiments and for identifying which step in an experiment may have failed.

Remember the following points:

- Have fun with the experiment. Experiments are an opportunity to learn, not an intelligence test. Everyone makes mistakes. Smart people learn from them.

- Careful planning ensures that the experiment can be done easily and safely. The use of controls simplifies data analysis. Be certain that the experiment is testing only one thing at a time.

- A good laboratory notebook aids in both the performance of the experiment and the analysis of the data. Advance planning helps ensure that the experiment will proceed smoothly. Discovering that vital information was omitted from the laboratory notebook is disappointing. Making that discovery on the night before the science fair is devastating. Review the notebook and analyze the data while there is still time to repeat the experiment.

- A failed experiment can be fixed. Positive controls help identify flaws in experimental design.

- Good laboratory technique is an asset, but mistakes will occur. Repeating an experiment provides an opportunity to improve it.

- There will be variability in the data from every experiment. Glitches are okay. The experiment may simply have to be repeated with more trials or samples.

- The data are telling you something. If the data prove that the model is incorrect, then further research may be necessary. If the experiment did not "work," there must be a reason. Try brainstorming to come up with possible explanations and alternative experiments.

NOTE

1   Prepare a blank table before starting the experiment. List all of the experimental treatments, and leave ample space to record the data. The experimenters can simply fill in the blanks as they do the experiment.

READING 6

SCIENCE, TECHNOLOGY, AND SOCIETY

By Ronald V. Morris

INTRODUCTION

The integration of science and social studies is natural. Many of the global environmental issues the people of the world face today have implications for economic, geographic, and public policy issues. The lives of people changed by the application of technology is certainly history. What did it mean for an Islamic cleric to discover calculus or the ability to bisect an arc? No longer must a sailor hug the coast of the Arabian Sea, hoping against hope that a storm will not blow him off course to be lost forever in the vast ocean. Now by using calculus and the sun and the stars, he can determine his position on the globe and find his way to safety. The consideration of the implications of those technological changes certainly called for decision making on the part of citizens. What does it mean that the highest elevation in the county is the local landfill? Some of the implications of new technology on people in less industrially developed portions of the world call for consideration as to the ethical implications of applying or imposing the technology on different ethnic or regional groups. The resulting social justice considerations certainly merit discussion and further examination.

When students go into the field to determine how minerals are mined and forests are logged, they reach their own conclusions about the wisdom of using these products. They gather data in the field and bring it to the classroom where they can continue their analysis by using print and electronic resources. They discuss what they can do to protect the environment through legislation, cleanup efforts, lobbying, and personal use of resources. They discuss the roles of government, media, and consumer groups in guarding and harvesting natural resources. They also determine the merits of recycling, reducing, and reusing materials made from natural resources.

Students need to explore the closed areas of society in order to understand the impact of technology. Even among themselves, adults do not usually talk about where things go once they enter a sewer; yet this closed and hidden world offers many adventures in discussing public policy. When considering public policy issues, students need to make decisions while considering the needs of society to determine the associated costs of implementing a new technology. Technology is never free; there is always a cost to society in closing as well as opening opportunities for the people who use, do not use, or misuse the technology. When technology speeds up a process or eliminates people from an economic process, the result requires that persons must quickly learn a new way to create income or obtain food, or both.

As students explore the impact of technology, science, and the environment, their explorations can be adventures, challenges, and adversity. All three of these form the foundation of memory. Students find physical experience a form of challenge when they walk distances to explore the site or experience primitive conditions on the site. Students find natural landmarks an adversity when they investigate muddy and slippery slopes when the elements do not cooperate. Students find cultural resource adventures when they find unexpected ideas and surprising sites.

When planning a study travel experience, it is important to remember that natural features help students understand the characteristics of a place (Halocha, 2005; Heines, Piechura-Coulture, Roberts, & Roberts, 2003; Howat, 2007). The story of a place can tie into many aspects of social studies as well as tying individuals to a place or community. These types of adventures with individuals, community members, and natural features allow students to have direct experiences with geography and even get dirty at the bottom of a sinkhole. Students scramble over coal strip-mine spoil looking for evidence of fern leaf fossils in the rock, look for fossils exposed by a road cut, or visit a deep limestone quarry. Unusual adventures such as these really interest students because they like both adversity and challenge. Scrambling down and then back up a steep hill to get to and from the base of a notable rock outcropping, walking into a dark cave at a state park with a flashlight, taking a cave tour when the lights go out, and taking a boat ride into a water-filled cave are all far better adventures than media experiences because they are real. Students store this information as part of their memories of the community (Bisbee, Kubina, & Birkenstock, 2000)—memories students later recount with other members of the community who shared the experience as a common bond or connection.

Some aspects of the past, such as the glaciations of an area, cannot be directly witnessed, making it difficult for students to experience these firsthand. The glaciers are gone, but the students use evidence of their presence to find out how glaciers changed the land. Students find out about glacial action by visiting nature preserves, which demonstrate the power of the glaciers to shape the land. If water or mud will be encountered, a change of footwear or boots is needed because student preparation is important in insuring successful and interactive adventures. If students will be in cold or wet weather or will take long walks, proper clothing and footgear is suggested; walking more than a mile especially in warm or cool conditions will seem long to suburban students, who are often not used to walking very far.

Natural landmarks have attracted and fascinated visitors for many years; they have a history all of their own (Aleixandre & Rodriguez, 2001). Meetings for peace or council many times occurred at definable geographic sites that can provide modern visitors space for picnic lunches. The distinctive characteristics of a place that attracted people in the past continue to attract people and will continue to attract people to the site.

Students prepare for the trip by examining newspaper stories about ground water from three different local papers. Students determine what impact agriculture, industry, and residences have on the quality of ground water. Students create a data retrieval chart with agricultural, industrial, and residential impacts listed on one axis and the dangers identified in the newspaper articles on the other axis. These might include runoff of pesticides, animal wastes, herbicides and fertilizers, dumping of heated water, metals, chemical wastes, detergents, sewer, and storm sewer water. They then fill in the cells created by the intersection of these ideas with information found from the newspaper stories.

Students then turn to print and electronic sources to determine what the effects of these pollutants are on humans. They interview environmental protection agents to determine what has been done to limit the dumping of these compounds in the past, and they interview university biologists to determine what more needs to be done. Students talk to a journalist to determine whether the laws are being enforced, and they talk to business people to determine how much it would cost to clean up these problems. Furthermore, they determine whether there is money to be made in the reclamation of discarded products by interviewing an environmental engineer connected with a landfill. Landfills demonstrate that recovering by-products of waste can generate income for the landfill company.

SCIENCE

State parks frequently contain cultural resources that may or may not be interpreted. Oftentimes park officials are conscious of the cultural resources in the park, but for a variety of reasons, including a lack of funds for interpretation or the attempt to protect the site from vandalism, they choose not to interpret a site. Other sites, however, definitely work to interpret their human history. Spring Mill State Park near Mitchell in south central Indiana has tree-covered land with springs, woods, sink holes, hills, and valleys; the park also contains a restored village complete with gristmill. Students learn about commercial and social interactions between people from this site. A group of fifth-grade students in Indiana explored the restored pioneer village in Spring Mill State Park. This isolated pioneer village attempted to be self-sufficient, so students made a list of all of the products created within the village (Morrell, 2003; Wolff & Wirmer, 2009). Students went from site to site examining tools and the contents of houses, barns, outbuildings, shops, and businesses to see what pioneer residents created in this community. When they returned to school, they compiled a list of the goods and services created by the community. The students found evidence of some obvious activities such as grinding grain into meal and sawing logs into boards at the mills, but evidence of other kinds of work such as preserving green beans by drying them and making furniture was more difficult to perceive. In this example, the fifth-grade students realized that the commonwealth of the community depended on the contributions of individual members. Natural features found in the park such as the water flowing out of the cave that was harnessed for the grist and lumber mills have significant stories of human interactions. The story of the twentieth century is illustrated in the prior use of the park land as homestead, farm, business, or cemetery. How the park was preserved reflects the ideals of the Progressive Era, and the site of the old Civilian Conservation Corps (CCC) camps illustrates how the Great Depression caused Franklin Delano Roosevelt to create work programs that ameliorated the suffering among unemployed young men.

Students have experiences that remind them of the past in addition to having current experiences (Wolff & Wirmer, 2009; Polette, 2008; James, 2006; Schulte, 2005; Dils, 2000). Students spent the night at the site of an old 1930s CCC camp, and they hiked trails that were laid out in the area of the CCC camp. Even though the students did not stay in real CCC barracks nor live like the CCC, and the trails were neither laid out nor followed by the CCC, by doing similar things to what the CCC did, the students learned, remembered, and commemorated the CCC.

TECHNOLOGY

When traveling to a military base, a space center, or a local space memorial, the teacher produces a first-person presentation to help students learn about lives of the first seven astronauts. The presentation helps students understand the beginnings of the space age. Figure 6.1 is a sample content outline for a first-person program covering Gus Grissom's life, how a multi-stage rocket works, and the Mercury, Gemini, and Apollo projects. Because the teacher is traveling, costuming is kept to a minimum—an old white dinner jacket is the only item required. The outline provides a rough frame for organization and content notes, but elaboration and questions can be fielded from student questions during and after the presentation. The first-person presentation provides another way for students to get a reinforcing experience with the content of the site.

Institutions play an important role in providing services to individuals and groups in communities; one group of institutions that reflects the idea of the connection of science, technology, and environment is medical history museums. Although the scope of their collections interprets the story of how medicine became a science by gathering data, they also provide visitors with interpretation about medical education and mental health care. Students see human specimens and photographic records. Students learn how individuals created institutions to care for people, conducted research, and tried to improve patient care through scientific processes. The story of citizens looking to science to help with medical and psychological problems reflects the practice of institutions trying to create specialized communities to assist with social problems (Bruce & Lin, 2009; Jornell, 2009; Lim, 2008). Students also see that the government can create institutions or agencies to help solve community problems as well.

Many early-elementary-aged classes seem to take field trips to working farms during the course of the year; they look at the animals and plants and find out where the food supply starts. It is a great place to start talking about all of the steps that our food goes through before it appears on our table. A farm is a great place to talk about soil and conservation, suburban sprawl taking over farmland, the economics of farming, and the future of farming. Even the ubiquitous trip to the farm can help students see the connections between science, technology, and environment in their own world. As students take field trips, they think about issues of sustainability and how they work a planet that is green when they are adults (Knapp & Barrie, 2001).

I. "How does a multiple stage Redstone rocket work"?
    A. Three stages
    B. Nose cone
        1. One person in the tip
        2. Lift off
            a. Shaking
            b. All the rides at an amusement park put together
            c. Pressed into the seat
    C. Each stage falls into the sea
    D. Heat shield
        1. Red hot
            a. Hot gasses glow past window
            b. Hot inside
    E. Parachutes deploy
    F. Capsule splash down in the water of the ocean
    G. Flotation balloons inflate
    H. Capsule picked up by helicopter
        1. Dropped on the aircraft carrier deck

II. Background of Gus Grissom
    A. Born in 1926
    B. Mitchell Public School education
    C. Waterwheel in Spring Mill fascinated him at an early age

III. Flying
    A. He sold his gun for his first airplane ride.
    B. In WWII he was a fighter pilot.
    C. He was a Purdue graduate in aeronautical engineering
    D. In Korea he was a fighter pilot.
    E. After the war, he became a test pilot.

IV. Project Mercury
    A. Original seven astronauts
    B. Purpose: Put man in space and return him safely
    C. Fifteen-minute flight
    D. Hatch
    E. Exploding bolts
    F. Nearly sank in the ocean

V. Gemini Project
    A. Twins: Two people in the capsule
    B. Purpose: Dock in space
    C. Meet in space
    D. Three sub orbits
    E. Unsinkable Molly Brown

VI. Apollo I
    A. Purpose: First man on moon
    B. Oxygen flash fire on the launching pad

FIGURE 6.1: Gus Grissom first-person presentation outline.

ENVIRONMENT

In the twenty-first century where technology envelops us, the natural world is divorced from us by contrived ventilation systems and unpleasant sights and smells are removed from our daily experience, it is important to show students where our conveniences originate (Figure 6.2). A favorite field trip is to the water treatment plant; the water flows straight from the reservoir with all of the remains of weekend boaters, swimmers, trash, leaf litter, and bird feathers (Pressick-Kilborn, 2009; Thomas, 2002). As the water passes through screens and then settlement tanks, the material that floats to the top is skimmed off, the material that falls to the bottom is removed, the water is shot into the air, and microbes eat their fill in the agitated water; then the water passes through sand and charcoal, and chlorine is added to the water. The students see all of these steps, including the lab where the tests are done; at the very end there is a drinking fountain. Most people do not feel thirsty at that point, but the teacher knows the students are really thinking about water quality—maybe for the first time in their lives.

- Airports
- Bird sanctuaries
- Caves
- Coking plants
- Dams
- Electrical plants
- Factories
- Farms
- Food processing plants
- Forests
- Lakes
- Landfills
- Medical history museums
- Military bases
- Mines
- National, state, or community parks
- Natural monuments
- Nature preserves
- Pharmaceutical museums
- Ponds
- Rivers
- Quarries
- Rock formations
- Sewage treatment plants
- Space centers
- Steam plants
- Reservoirs
- Water treatment plants
- Wet lands

FIGURE 6.2: Science, technology, and society field trip locations.

The trip to the local landfill is not quite so much fun, but it is interesting to see how everything is sorted. Truck after truck comes in all day bringing waste; the staggering amounts are fascinating, but the amount that can be recycled is also amazing. Every possible effort is made to use all the products of waste. Human interactions with the environment are natural and needed parts of the exploratory nature of study travel experiences. If students take study travel trips to cities, then they need to consider the environmental implications of the site (Figure 6.3).

It is important to consider the costs by working on-site to gather data on the costs, witnessing the impact on the land, and documenting the costs through drawings. For elementary school students drawings serve as a form of note taking while they are learning in field experiences. Students can record their observations, reflections, and recommendations by creating quick pencil sketches of what they are discerning. Through their note taking, they bring ideas and information back to the classroom, where further discussion occurs. Students make detailed drawings about one site or they compare several sites through their drawings.

Students compare sink holes and springs when studying ground water pollution in a karst region. The students hike along the Lost River to record sketches of swallow holes, a karst window, and the rise of the river. The students place captions on their drawings showing the pollutants that impact each of the areas. Back in the classroom the students compile all the different pollutants that flow into the cave system and compare those pollutants with those found in the local streams. After this they determine whether legislation is needed to protect ground water that is used for public or private drinking supplies. When students return from the field, they need to create newspaper stories for their local newspaper and their school newspaper describing what they found. The results of these news releases also serve as assessments and are graded on a rubric (Figure 6.4).

> I. Culture
> II. Time, continuity, and change
> III. People, places, and environments
> IV. Individual development and identity
> V. Individuals, groups, and institutions
> VI. Power, authority, and governance
> VII. Production, distribution, and consumption
> VIII. Science, technology, and society
> IX. Global connections
> X. Civic ideals and practices

FIGURE 6.3: NCSS (1994) national standards.

Natural Resource Development
4 Documented two (2) achievements and two (2) failures in extracting the natural resource
3 Documented both an achievement and a failure in extracting the natural resource
2 Documented an achievement or a failure in extracting the natural resource
1 Documented no achievements or failures

Community Resource People
4 Quoted two or more resource people showing opposite sides of an issue
3 Quoted two resource people
2 Quoted one resource person
1 Quoted no resource people

Advocacy
4 Asked for the reader to take action
3 Took a position that encourages social justice
2 Identified a position that is socially responsible
1 Took no position

Controversial Issue
4 Described both sides of the issue
3 Identified both sides of the issue
2 Identified a controversial issue
1 Identified no controversial issue

Responsibility
4 Encouraged a balanced and fair accounting of both sides of the issue
3 Respectfully acknowledged differing views on the topic
2 Respectfully illustrated only one side of an issue
1 Took no responsibility

FIGURE 6.4: Assessment.

WORLD CONNECTIONS THROUGH TRANSPORTATION

What is the role of government in encouraging good transportation? The government plays a variety of roles: from regulation of transportation systems, to creating and maintaining them with tax funds, to taxing them for revenue, to ensuring their safety. Citizens need to understand how these transportation networks function and how the government interacts within those systems. Citizens need public policy that encourages transportation and trade while using a responsible balance of producing revenue to pay for government expenditures for transportation systems. Citizens need safe efficient transportation that helps them move goods from one part of the nation to another.

Citizens need to understand the role the government plays in keeping them safe on the water, air, and land through vehicle design, regulation, and enforcement.

The continued development of transportation corridors may have included the paths of Native People or the traces of migrating North American bison. These paths such as modern US 150 still function after years of upgrading by pioneers, private toll roads, or state highway workers. The features found along the way include residences, factories, schools, churches, businesses, and courthouses. These sites build on top of previous structures or stand as neighbors to co-existing structures spanning two hundred years of building practices. When they are in the field, students study the built environment, which includes all of these examples along the route.

People have developed a number of ways to connect with other people through trade and communication. At first these methods were slow, but over time they have become faster and more reliable in linking people to their world. Students examine a number of these different methods of communication and transportation then they compare them with the communication and transportation systems with which they are currently familiar (Lybolt, Techmanski, & Gottfred, 2007; Maher, 2004; Alleman & Brophy, 2003A, 2003B).

The people whom students meet when they travel to learn about transportation are often as interesting as the sites the students visit. Students who have played with toy cars, buses, boats, airplanes, and trains such as Thomas the Tank Engine easily assimilate the new information about transportation networks that they learn by taking field trips and apply it to the modern world of distribution systems.

One way to set up a field trip is to organize it around a transportation corridor such as a canal, railroad, river, or road. Each stop along the route can be connected to the history of the corridor, thus giving a theme and coherence to what might otherwise appear be just a jumble of stops. This sort of field trip can provide a theme that easily goes beyond any one time period; students therefore see a variety of sites juxtaposed without a specific historic context. Students experience sites that may not normally be open for tourists; these sites may have little or no cost, but it is not likely that any interpretation will be provided. The transportation corridor is easy for students to map; they chart their way through the field trip both prior to the event and during the event.

Internal Improvements

Students need to create maps of transportation systems prior to field investigation of the transportation systems. Whether these are maps of roads, harbors, canals, rivers, or rails, each transportation system intersects with other forms of transportation. The students examine points of embarkment and disembarkment, towns along the way, and where other transportation systems cross the transportation system. The students find how traffic, trade, and communication get from one community to other points around the nation or the world. Furthermore, students learn about points where cargos are moved from one type of transportation to another.

Students trace the National Road across the capital cities of the then western states of Ohio, Indiana, and Illinois. Running from Cumberland, Maryland, to Vandalia, Illinois, the road was the obvious attempt to connect the western states to the east coast. Immigrants took the road to enter the area, farmers moved their goods to market, travelers carried news and mail, and traders prospered all along its route. It was updated to US 40. All along the route towns sprang up and at places such as the Whitewater River or the Whitewater Canal major communities developed because of the intersection of trade and commerce.

In east central Indiana students follow the National Road (Figures 6.5 and 6.6), which was the first federally funded highway. The National Road was designed to link Cumberland, Maryland, in the east, with the west, by going through the three state capitals of Columbus, Ohio; Indianapolis, Indiana; and Vandalia, Illinois. One of the best stops along this route is the Huddleston Farm House Museum, which is operated by the Indiana Landmarks. This group is devoted to working for the preservation of historic buildings and regional landscapes, and as part of that mission they have restored and operated the structure as a museum to interpret life on the National Road and the westward migration as well as incorporating the best practices of preservation (Ferretti, MacArthur, & Okolo, 2007; Clark, 2000). The students' next stop is the home of Levi Coffin, the president of the Underground Railroad. This site reflects how the migration of Quakers, who left slave states for reasons of conscience, intersected the National Road and how this promoted the Underground Railroad activity in the area. Students later stop at the junction of the east fork of the Whitewater River. The falling water at this juncture with the National Road provided mechanical energy for the emerging industries.

8:00 a.m.	Depart from school
9:30 a.m.	Gaar House
10:30 a.m.	Depart
11:15 a.m.	Levi Coffin Home
12:15 p.m.	Depart
1:00 p.m.	Lunch
1:30 p.m.	Depart
2:00 p.m.	Wayne County Historical Museum
3:00 p.m.	Depart
3:30 p.m.	Huddleston Home
5:00 p.m.	Depart
6:00 p.m.	Return to school

FIGURE 6.5: National road tour.

Gaar House	$2.00 per person
Levi Coffin Home	$.25 child, $1.00 adult
Wayne County Historical Museum	$.75 child, $1.50 adult
Huddleston Home	$1.50 child
Total per child	$4.50

FIGURE 6.6: Fees for the national road tour.

The students can see the home of one of the industrial families when they stop at the Gaar Mansion. Their next stop is the Wayne County Historical Museum, where they learn how Mrs. Julia Gaar traveled throughout the world and brought back curiosities to display first in her home and then later placed them in the museum to interpret the world to her neighbors. Her family shipped their products by using transportation networks, and she traveled the world by using transportation networks. Both of these examples help the students to learn about the role of movement in geography and history.

The canal system, which laborers dug in the early 1800s, has been erased from the landscape in many places, but in certain places it can still be seen by careful observation. Students visit Indiana's Canal Park and Interpretive Center at Delphi, Indiana, which affords an interesting interactive history of the Wabash and Erie Canal through the use of working models and recreated canal features. The students also climb into the remains of an old lock, walk in the old canal and beside the restored canal on the tow path, walk under the canal where it crosses local streams, and walk over a river where the canal crossed it. Some of this exploration requires students to go down a muddy bank and get their feet wet as they look for clues of the existence of former canals. On field trips students look for clues and examples of places where the physical landscape changed over time.

The docents of the Howard Steamboat Museum show a different type of transportation system in Jeffersonville, Indiana. The first floor of the museum interprets the life and times of the Howard family when they built boats for the Ohio, Mississippi, and Missouri Rivers; the second floor portrays the history, geography, and economics of the family-owned and operated company and is interpreted with models, artifacts, photographs, and documents. Floating palaces and work horse boats all powered by steam engines were the specialty of the family; in addition to producing a large number of boats, the Howard family produced some of the best known boats on the river system. Their company continued to make boats for World War II, and boats continue to be made at the site where their company once stood. For students who have not grown up on a river, the river itself provides a variety of new sensory experiences to help them think about how the river has been and is presently being used by people.

While students are on the river, they see an international harbor where boats bring in raw materials and pick up agricultural and manufactured products to exchange on a world market. The boats, containers, railroads, cranes, and warehouses are on a super-human scale, and all of these physical symbols of international trade are available for student inspection. Similarly, all of the raw materials of trade are visible on the wharves while being moved to and from warehouses. Usually transportation systems intersect, which mean that railroad corridors, roads, wharves, and air hubs cross one another to provide ways for goods to be transported to the next location. Students see how goods and services travel from one transportation corridor to another, and they learn how raw materials and finished goods move in a world economy.

Students bring notes back from the field by using a sheet of paper and the side of a crayon with the paper removed. By rubbing the side of the crayon over the paper held steady on an incised stone or raised metal plate, students bring images back to the classroom as a form of note taking. Students make rubbings of cast historical markers, information plates on modes of transportation such as trains or steamboats, or they rub informational text from stones. They hang these rubbings on the walls as a reminder of the trip or they use the information found in them to try to acquire more information about individuals, groups, or institutions from the past with the help of additional print or electronic sources. They find primary source documents, illustrations, and photographs to determine what other stories they can find about the site that are not told in the text.

Students determine controversial issues that surrounded the construction or improvements of these transportation routes. Students determine that not everyone thought that they were a good idea and that not everyone approved of them. Some people had their towns bypassed by the transportation routes; others had their farmland crossed, thus losing easy access to their land; still others had their land seized through eminent domain. Students see that people in the past as well as in the present are not usually all of one mind; there was dissent and faction in the past just as there is in the present. Modern infrastructure projects, including roads, light rail, air hub, and harbors, displace and disrupt people in addition to bringing new business and connections to an area.

When using primary sources, students find a variety of information they use to create information files for students in the class next door (Figure 6.7).

EXPLORING DIVERSITY OF THOUGHT AND COMMUNAL GROUPS

The frontier can be misperceived as colorless, genderless, and filled with lusty pioneers, who were all the same and held no beliefs while conquering the wilderness. Excellent teachers help students see diversity in thought, country of origin, color, age, and gender. The people who made up the group called pioneers came from a variety of countries and states. They brought a variety of religions or none, and they had passionate beliefs about how government ought to operate. These are the people who fill the pages of the history books, and they need to have their multiple thoughts represented in social studies classrooms.

Careers
4 Identified two transportation related jobs that require a college degree
3 Identified a transportation related job that requires a college degree
2 Identified a transportation related job
1 Identified no job

Sources
4 Used more than one primary and more than one secondary source
3 Used one primary source
2 Used one secondary source
1 Used no sources

Transportation Policy
4 Identified a transportation policy, cited sources to justify it, and discussed its implications
3 Identified a transportation policy and cited sources to justify it
2 Identified a transportation policy and justified it with a rationale
1 Identified no transportation policy

Community
4 Provided more than one primary or secondary source for their peers
3 Provided information for peers
2 Provided information for other students in another room
1 Provided no information

FIGURE 6.7: Information file assessment.

Good primary and secondary sources exist to illustrate the diversity of thought represented by these people. Fortunately, there are still some sites that tell the stories of these people. Some of these sites are opened by ecclesiastical hands to visitors, and other sites are opened by educational foundations that are eager to educate visitors. These sites are difficult to interpret because bricks and mortar only tell so much; the unfortunate habit of *presentism* often occurs, whereby visitors think that the people acted in the past with today's motives and conventions. Museum educators need careful interpretation to tell the story of religious dissenters in the face of a romanticized enthusiasm for the past.

Communal societies from both the past and the present are difficult to understand. One of the most difficult parts of telling these stories is that critics and apologists for the groups created primary sources at the time. In their beginnings the groups were controversial, over time they became no less so. Of course fear, disdain, and prejudice of people who were different in very important ways—such as their means of establishing family and community—are still controversial issues. Even the cloistered communities of convents and monasteries are still clouded in jokes and ignorance.

Teachers usually teach the pioneer period as a festival of log cabins, food, split rail fences, and fun, but the frontier had significant diversity of thought and unique groups of people. Teachers help students explore these varying viewpoints by finding discrepancies in the idea that all pioneers were rugged individualists or even that all pioneers lived on farms in the wilderness. At the edge of civilization, some groups of people lived surprisingly comfortable lives in well-established communities. They lived in urban communities with all of the advantages of Europe or Philadelphia within the towns they created in the wilderness. Some of these utopias were religious and others secular, but the members of each group believed that they could create a community for a specific purpose that would change the world forever.

One such group was the celibate, communalistic, and millinenialistic followers of George Rapp of Harmony on the Wabash; the world called these German immigrants The Harmony Society or Harmonists. They carved a quality of life out of the wilderness equal in convenience to living in Philadelphia or Boston and produced manufactured goods equal to those produced in Europe. While the Lincoln family was living in a rough cabin with a dirt floor just a few miles to the west, the Harmonists enjoyed brick buildings with glass windows, musical concerts, and landscaped gardens. This group, like many others such as the Amana, Moravians, Oneida, Shaker, and Zoar, made contributions to the opening of the west from their communities. The contributions and thoughts of these people represent an important alternative path of study about those who settled in the west as pioneers (Peterson, 2004; Levstik & Groth, 2002; McMillan, 2000).

Today, Historic New Harmony is an open-air museum preserving much of the residential portion of the town. As with most open-air museums, there is much that must be left to the imagination that a modern visitor cannot see nor experience. When students visit the town, the most obvious things missing are the two massive churches that dominated the center of town. They must also imagine the dirt streets lined with Lombardy poplar trees, the gardens filled with flowers, and the music that visitors described hearing at a variety of times throughout the day. The extensive fields and herds are difficult for open-air museums to interpret for visitors, and the large factories and keelboat construction at the side of the Wabash River are prohibitively expensive to replicate. Today the steady stream of outsiders, who come to trade and observe contrasts with a time in the past when people knew everybody in the community.

Students who come to Harmony on the Wabash now can try to solve two puzzles. First at Harmony, Pennsylvania, and later at Economy, Pennsylvania, the Harmony Society created large stone sinks that drained water through the wall and out of the building. Why are none of these built in Harmony on the Wabash? The students need to create three hypotheses, determine which one they think is most feasible, and state the reasons they believe this one is correct. The second puzzle is that at both sites in Pennsylvania the Harmony Society had extensive wine cellars, but these do not seem to exist in the Indiana settlement. Once again the students need to determine three hypotheses, determine which one they think is most feasible, and state the reasons they believe it to be correct.

Students can perform structured role-play to simulate life in Harmony. Through this activity the students determine the types of jobs that were available in the town and the beliefs of the Rappites including holding property in common, equality of members, how to settle problems, living in non-violence, and religious beliefs (Figure 6.8). Students take on different roles as members of the town, perform physical mnemonics to remember ideas, and engage in open-ended questioning. The different roles help the students to understand the various types of labor conducted in the town, and by being active the students connect the kinesthetic motions with the ideas they need to remember. The open-ended questioning helps the students to understand what motivated the group and how they would need to deal with similar types of situations. The role-play also

---

I. Jobs
    A.   Farmers (farm symbol)
    B.   Mechanics (pull lever)
    C.   Traders (*push me pull you*)

II. Build Community
    A.   Common
         1.   What problems could be caused if everything were owned by the community and not by individuals?
    B.   Equal (*balanced hands*)
         1.   Free store (*cash register*)
              a.   How could you keep people equal if there were a free store?
              b.   What would the neighbors think of a free store where they had to pay money to trade?
    C.   Harmony
         1.   Who would settle problems?
    D.   Peace (*sign*)
         1.   What would the neighbors think about members of the community not serving in war?
         2.   What would the neighbors think about their liking the Indians?
         3.   By not serving in war, do you think that the community members were consistent with their beliefs?
    E.   Religion
         1.   Jesus was coming to Jerusalem
              a.   What would you want to do?
         2.   The world was going to end
         3.   No children (*international sign*)
              a.   Why would you not want to have children?
              b.   Do you think that the Harmony Society had a good or bad life without children?
    F.   After ten years, the Harmony Society left Indiana.
         1.   What were their jobs?
         2.   Where were their markets?
         3.   Would you have liked life in Harmony on the Wabash?

FIGURE 6.8: Role-playing the history of the Harmony Society on the Wabash.

> I. Gertrude lived in Harmony from age 7 to 17.
>   A. Granddaughter of Father Rapp
>   B. Much loved by the community
>   C. Sick: coughed until she choked
>   1. Sang and jumped around
>   D. Needlepoint
>   E. Flowers and gardens
>   F. Learned English from Shakers at West Union
>   1. She received letters from Shaker friends
>   2. She wrote her grandfather's letters for him in English

FIGURE 6.9: Gertrude Rapp: Child in Harmony.

describes and interprets how these people conducted business and moved goods to markets by making economic decisions while taking geographic locations into account.

Gertrude Rapp lived in the Harmony community as a child from age seven to seventeen. The primary sources of the Rapp community give the best description of her because she was the granddaughter of Father George Rapp (Figure 6.9). Students read these documents when they are on the site to find out what life for one child would have been like in the community. Students learn about health care, leisure activities for children, and the presence of other communal groups on the frontier from these documents. Students also discover how historians learn by using evidence from primary sources. By using the primary sources in the field, students can get a better idea about what happened in the life of people and especially that of children their own age.

Owen and New Harmony

When the Harmony Society left the Wabash and returned to Pennsylvania, they sold their entire community to Scottish textile manufacturer, philanthropist, and social reformer Robert Owen. The utopia he attempted was called New Harmony, and it was based on science and universal education. The group held all property in common, but they were neither religious nor celibate. The members of the community entertained ideas that were very foreign to those of their geographic neighbors including the equality of women and the equality of people regardless of the color of their skin. These ideas allow students to look at intellectual history, see how ideas emerge, evolve, and are tested, accepted, or rejected by society (Haibo, 2009; Milewski, 2008; Sevier, 2002).

Before, during, or after the trip, the students engage in structured role-play to simulate life in New Harmony with the Owenites and compare it with their life. Students take roles, act out physical mnemonics, and engage in open-ended questioning. Students consider the jobs of thinkers, scientists, and educators, and the equality of members of the community, and they contemplate why the Owenites were still unable to solve

> I. Jobs
>   A. Thinkers (thinker)
>   B. Scientists (pour test tubes)
>   1. Geology
>   2. Shells
>   3. Insects
>   4. Books
>   C. Educators (write)
>   1. School all of your life
>   a. What jobs do the Owenites not have?
>   b. What problem could this cause?
> II. Everyone was equal (balance hands)
>   A. Owned all in common
>   1. What problems could this cause?
>   B. Fights (act)
>   1. Who could settle problems?
>   C. Split town
>   D. Fights (act)
>   E. Quit common community
>   F. People stayed
>   1. Why did people stay?
>   2. What kind of people occupied New Harmony?
>   3. Were the people staying here going to benefit the community or cause a problem for it?

FIGURE 6.10: Role-playing the history of Robert Owen's New Harmony.

their problems (Figure 6.10). Students examine the reasons the community failed and speculate about the causes for the decline and failure of the community. Students also consider whether there are connections between this event and other events in the present or in the past. When they identify an event, they need to give examples of how it is and is not similar.

Students also learn about the site through a first-person presentation in the character of Joseph Neef, one of the educators who came to the town for the utopia. Children lived in boarding schools, and the stories of students from the past are both compelling and whimsical descriptions. The utopia failed, but the people stayed. Joseph describes the scholars he worked with, the trip to the community on the keel boat filled with so many thinkers and scientists that it was called the boat load of knowledge, how he ran the upper school with his wife and daughters, his career, and how he worked with other members of the community (Figure 6.11). Whether given on-site or prior to going to the site, the first-person presentation helps bring a human dimension to the story (Howes & Cruz, 2009; Morris, 2009: Cruz & Murthy, 2006). Students like to hear about the adventures of their peers even when they are separated by time.

> I. Searching for Victor Collin Duclos
> II. Two French students fell through the ice and nearly drowned when the Philanthropist stopped near Louisville.
> III. Teacher and conducted the upper school with my wife and daughters
>   A. Madame Fretageot has the lower school
>   B. Colleague of Pestalozzi
>   C. Taught for Mr. MaClure in Philadelphia
>   D. Quit teaching and moved to a farm near Louisville
> IV. Fought in the Napoleonic Wars
>   A. Favorite friend of Robert Owen
>   B. Kindly man
>   1. Scared students because he swore
>   C. Education
>   1. School
>   2. Trade
> V. Checked with Mr. Say

FIGURE 6.11: Joseph Neef first-person presentation outline.

Social World
4 Compared two communal groups including flaws, strengths, dangers, and promise including a group that still has members in the area
3 Compared two communal groups including flaws, strengths, dangers, and promise
2 Described one communal group including flaws, strengths, dangers, and promise
1 Described no communal groups

Controversy
4 Included cultural similarities and differences
3 Quoted appropriate sources for each group
2 Included two different points of view for each group
1 Included only one point of view

Literature
4 Cited both primary and secondary sources that offer conflicting perspectives about the two groups
3 Cited primary sources that offer conflicting perspectives about the two groups
2 Cited secondary sources that offer conflicting perspectives about the two groups
1 Included only one point of view

FIGURE 6.12: Assessment.

Probably the most sustainable celibate communal religious groups were established by Catholic orders at Ferdinand, Oldenburg, and St. Meinrad, in Indiana. In their prime these groups owned vast tracts of land, large herds, employed many of the surrounding parishioners, practiced large-scale agriculture, engaged in cottage industries, and, despite being mono-sex communities, functioned in a way surprisingly similar to West Union and Harmony on the Wabash. Their tendency to employ parishioners was the largest difference in contrast to other frontier deliberate communities; in this regard, they tended to look like European monasteries and convents. They divested themselves of most of their land in the early twentieth century, but it is still possible to visit the main buildings of each of these cloisters. The brothers and sisters are still there carrying out their holy and secular duties with connections to Europe, the traditions of their respective community, and their history at the site.

The students create an essay comparing communal groups that might include Owenites, Harmonists, Shakers, Benedictine Sisters at Ferdinand, or Benedictine Brothers at St. Meinrad (Figure 6.12).

REFERENCES

Aleixandre, M. P., & Rodriguez, R. L. (2001). Designing a field code: Environmental values in primary school. *Environmental Education Research, 7*(1), 5–22.

Alleman, J., & Brophy, J. (2003A). Comparing transportation units published in 1931 and 2002: What have we learned? *Journal of Curriculum and Supervision, 19*(1), 5–28.

Alleman, J., & Brophy, J. (2003B). History is alive: Teaching young children about changes over time. *Social Studies, 94*(3), 107–110.

Bisbee, G., Kubina, L., & Birkenstock, S. (2000). Create your own field guide. *Science Teacher, 67*(5), 42–43.

Bruce, B., & Lin, C. C. (2009). Voices of youth: Podcasting as a means for inquiry-based community engagement. *E-Learning, 6*(2), 230–241.

Clark, A. D. (2000). Living the past at Oak Hill School. *Now & Then, 17*(3), 13–17.

Cruz, B. C., & Murthy, S. A. (2006). Breathing life into history: Using role-playing to engage students. *Social Studies and the Young Learner, 19*(1), 4–8.

Dils, A. K. (2000). Using technology in a middle school social studies classroom. *International Journal of Social Education, 15*(1), 102–112.

Ferretti, R. P., MacArthur, C. A., & Okolo, C. M. (2007). Students' misconceptions about U.S. westward migration. *Journal of Learning Disabilities, 40*(2), 145–153.

Haibo, Y. (2009). Naxi intellectuals and ethnic identity. *Diaspora, Indigenous, and Minority Education, 3*(1), 21–31.

Halocha, J. (2005). Developing a research tool to enable children to voice their experiences and learning through fieldwork. *International Research in Geographical and Environmental Education, 14*(4), 348–355.

Heines, E. D., Piechura-Coulture, K., Roberts, D., & Roberts, J. (2003). PARKnerships are for all. *Science and Children, 41*(3), 25-29.

Howat, C. (2007). Le vieux carré: A field trip to a New Orleans parish. *Social Studies and the Young Learner, 20*(2), 24-29.

Howes, E. V., & Cruz, B. C. (2009). Role-playing in science education: An effective strategy for developing multiple perspectives. *Journal of Elementary Science Education, 21*(3), 33-46.

James, A. (2006). *Preschool success: Everything you need to know to help your child learn*. Indianapolis, IN : Jossey-Bass.

Jornell, W. (2009). Using YouTube to teach presidential election propaganda: Twelve representative videos. *Social Education, 73*(7), 325-329, 362-363.

Knapp, D., & Barrie, E. (2001). Content evaluation of an environmental science field trip. *Journal of Science Education and Technology, 10*(4), 351-357.

Levstik, L. S., & Groth, J. (2002). "Scary thing, being an eighth grader:" Exploring gender and sexuality in a middle school U.S. history unit. *Theory and Research in Social Education, 30*(2), 233-254.

Lim, C. P. (2008). Global citizenship education, school curriculum and games: Learning mathematics, English and science as a global citizen. *Computers & Education, 51*(3), 1073-1093.

Lybolt, J., Techmanski, K. E., & Gottfred, C. (2007). *Building language throughout the year: The preschool early literacy curriculum*. Baltimore, MD: Brookes.

Maher, R. (2004). "Workin' on the railroad": African American labor history. *Social Education, 68*(5), S4-S9.

McMillan, E. P. (2000). Traveling west in 1845: A first grade project. *Social Studies and the Young Learner, 13*(2), 28-31.

Milewski, P. (2008). "The little gray book": Pedagogy, discourse and rupture in 1937. *History of Education, 37*(1), 91-111.

Morrell, P. D. (2003). Cognitive impact of a grade school field trip. *Journal of Elementary Science Education, 15*(1), 27-36.

Morris, R. V. (2009). *Bringing history to life: First person presentations in elementary and middle school social studies*. New York, NY: Roman & Littlefield.

National Council for the Social Studies. (1994). *Expectations of excellence: Curriculum standards for social studies (Bulletin 89)*. Washington, DC: Author.

Peterson, C. (2004). *Jump back in time: A living history resource*. Portsmouth, NH: Teacher Ideas Press.

Polette, N. (2008). *Get up and move with nonfiction*. Portsmouth, NH: Teacher Ideas Press.

Pressick-Kilborn, K. (2009). Steps to fostering a learning community in the primary science classroom. *Teaching Science, 55*(1), 27-29.

Schulte, P. (2005). Social studies in motion: Learning with the whole person. *Social Studies and the Young Learner, 17*(4), 13-16.

Sevier, B. R. (2002). The creation and content of an early "multicultural" social studies textbook: Learning from "people of Denver." *Theory and Research in Social Education, 30*(1), 116–141.

Thomas, J. A. (2002). How deep is the water? *Science and Children, 40*(2), 28–32.

Wolff, A. L., & Wirmer, N. (2009). Shopping for mathematics in consumer town. *Young Children, 64*(3), 34–38.

SECTION II — QUESTIONS TO PONDER...

1. What are the science process skills?

2. What is an observation, and how is it different from an inference?

3. What are the parts of an experiment?

4. How is the 5E model different from other lesson plan models you have studied?

Section III

SCIENCE EDUCATION AND SOCIETAL FACTORS

Objectives

- Describe the relationship between science, technology, and society.
- Analyze important factors when differentiating science lessons and activities.
- Identify why it is important to address social justice while teaching science.

Key Concepts

differentiation: The manipulation of lessons to address the needs and abilities of learners

social justice: A broad political and philosophical concept which holds that all people should have equal access to wealth, health, well-being, justice, and opportunity

Introduction

This section of the book addresses societal ideas important to teaching science. It examines the relationship between science, technology, and society (STS). This relationship is important to consider in order to understand that science does not happen in a void but is used to create technology, and it has an impact on society. Society will use technology for science, and science will have an impact on technology.

Differentiating science lessons and activities is an important tool in meeting students' learning needs. These changes can be based on learning styles, abilities, cognitive development, or interests. Being able to differentiate lessons demonstrates a desire to ensure that your students will learn and succeed.

Social justice is a concept that goes beyond multicultural education. In science, this is important to address in order to ensure the inclusion of women and people of color in STEMJ careers.

READING 7

DIFFERENTIATING SCIENCE PEDAGOGY

By Anila Asghar

STUDENTS' INTUITIVE IDEAS ABOUT THE NATURAL WORLD

A pervasive assumption that has influenced the beliefs and practice of many, if not all, science educators, is that the formal science education in school is the main source of scientific ideas and models about the natural world. But as Driver, Squires, and Wood-Robinson (1994) state, "learning about the world does not take place in a social vacuum" (p. 3). Science educators must recognize that students' interactions and experiences with the natural world shape their ideas in significant ways. Often, science teachers exacerbate the situation by viewing each student as a *tabula rasa* to fill with principles and theories, or they presume perfect prior knowledge with which to build more-complex concepts upon. The possibility of students' intuitive ideas about natural phenomena is rarely acknowledged. Teachers must realize that children "don't just passively receive information," but instead "operate on it and transform it," (Baker & Piburn, 1997, p. 31) based on global and personal experiences. Driver (1985) contends that children's perceptual experiences influence their conceptual frameworks around scientific models. Sadler and colleagues (Schneps & Sadler, 1997) conducted a thought-provoking study with school children and university graduates to elucidate how young people, from elementary school through college, perceive the relationship between scientific principles and the world around them. The interviews are captured in a video titled "Minds of Our Own—Lessons From Thin Air." The video shows a thread linking the *same* general misconceptions about basic science concepts, such as photosynthesis, light, states of matter, and seasons, from elementary students to graduating science and engineering university students. Even after years of top-notch schooling, these graduates' basic ideas about nature resonated with the ones held by elementary and middle graders.

The most important lesson for science educators, curriculum developers, and researchers is that many students do not develop a clear and robust understanding of the fundamental concepts in science. This challenges the assumption that science instruction can fill children's minds with science models that they will carry in their heads all their lives. Research in science education (Asghar & Libarkin, 2010; Fuson, 1988; Harlen, 1987; McCloskey, 1983; Smith, Maclin, Grosslight, & Davis, 1997), on the contrary, shows that students' experiences with natural phenomena and their everyday language heavily influence their understanding of science conceptual frameworks. According to McCloskey (1983), people develop remarkably well-articulated alternative conceptions on the basis of their everyday experience, and they are deep-seated and persist even after exposure to formal science instruction. When scientific models are presented to them, they interact with their preexisting notions in interesting ways. Their intuitive skills are integrated with the new information, often producing localized and incoherent frameworks, which they apply in an inconsistent manner to explain and solve problems. The *abstract and formal* language of science could exist alongside children's intuitive frameworks.

Scholars agree that children's ideas about the world "differ in conceptual content from those of scientists" (Brewer & Samarapungavan, 1991; Driver, 1981; Driver et al., 1994; Harris, 1994; McCloskey & Kargon, 1988; Thagard, 1989; Vosniadou, 2007; Vosniadou & Brewer, 1992; Wiser, 1988). Nevertheless, there is a continuing debate over the nature of children's ideas. Some researchers argue that children's intuitive knowledge is fragmented, unstable, and consists of weakly connected systems of ideas (Claxton, 1993; diSessa, 1983, 1993; Vosniadou, 1994). Other researchers contend that children's novice theories or "explanatory frameworks" exhibit some properties that are similar to scientific theories (Brewer & Samarapungavan, 1991; Nakhleh & Samarapungavan, 1999; Samarapungavan & Wiers, 1997; Wiser, 1988) that children may use to make predictions, develop cause-and-effect explanations, and apprehend and solve scientific problems (Samarapungavan & Wiers, 1997). These intuitive frameworks interact with the new concepts that students learn in science classes in interesting ways, often making it difficult for children to develop expert knowledge.

Research on K–12 and college students' understanding of scientific models suggests that alternative frameworks about fundamental physical, geological, and biological concepts may persist despite years of exposure to science instruction in formal education settings (Asghar, Libarkin, & Crocektt, 2001; Driver, 1985; Driver & Russell, 1982; Osborne, 1980; Perkins, 1992). Children's underlying preconceptions are resistant to change, and they influence their learning of new scientific ideas in important ways (Asghar & Libarkin, 2010; Libarkin & Asghar, 2002). Asghar and Libarkin, for instance, found that college students carried several underlying misconceptions about the concepts of force, velocity, acceleration, and gravity. Moreover, their ideas about gravity were strikingly similar to elementary students' intuitive conceptions of gravity.

CHILDREN'S ALTERNATIVE CONCEPTIONS OF CONSERVATION OF MASS

Findings from a study focusing on children's alternative frameworks related to the conservation of mass are discussed herein to illustrate the ways in which they interfere with their developing understandings of this concept. Scientists view matter as composed of indestructible particles (atoms) that are unique to and characteristic of each element. The total number of atoms in a system is conserved during every physical or chemical transformation. By accounting for these particles, chemists make accurate predictions about the physical world. For example, scientists can calculate the optimal ratios of chemicals needed for certain reactions so that no components remain uncombined.

Research shows that a considerable proportion of students hold a fairly stable model of matter in which matter is perceived as "continuous and static." (Children think that a substance is composed of matter that can be infinitely divided into small pieces of the same material—a non-particulate view. Water and steam are viewed as having entirely different compositions.) This model influences the learning of the "abstract" particle model of matter in school (Driver, 1985, p. 36). Researchers have found that children (9-13 years) possess inconsistent frameworks about matter across a range of substances from "continuous solids to particulate solids" (Nakhleh & Samarpungavan, 1999; Driver, 1985). Piaget and Inhelder (1974) state that children's perceptual and concrete experiences influence their reasoning. Children tend to make sense of a substance by observing their macroscopic properties, and if a substance disappears during a physical or chemical change, it is no longer the same.

Most 7- to 13-year-old children hold a continuous view of matter (Briggs & Holding, 1986). Qualitative studies reveal the range of children's non-particulate ideas. Children think, for instance, that a solid is either lost completely or loses weight while dissolving and melting (Driver et al., 1985; Stavy & Stachel, 1984). Children possess interesting preconceptions around combustion. They tend to think that metals would become lighter after burning because "certain things" are "burnt away" (Driver, 1985, p. 38).

Children also possess diverse ideas about the dissolving of sugar and other substances in water at various stages of development. Children up to age 8 describe the dissolving of sugar as something that "just goes," "disappears," "melts away," "dissolves away," or "turns into water" (Driver, Squires, Rushworth, & Wood-Robinson, 1994). Older students (12-16 years) think that sugar "goes into tiny little bits" as it dissolves or "mixes with water molecules" (Abraham & Williamson, 1994; Bar, 1986; Briggs & Holding, 1986; Driver et al, 1994; Nussbaum, 1985). Another frequently expressed idea is that a solid loses weight when it changes into liquid state (Stavy & Stachel, 1984). Some children think that sugar is "up in the water" and is not "pressing down" on the bottom of the vessel (Abraham & Williamson, 1994). Some students think that the sugar may still be present in water after dissolving but is "lighter." A plausible explanation for this reasoning is that after dissolving in water, sugar is not a solid, its mass therefore cannot act on a surface in the same way (Driver, 1985).

For some children, a solution is a "single substance" rather than a mixture of different substances. The non-particulate nature of a solution is more prevalent among 10-12 year olds (Briggs & Holding, 1986). When students' (age 9-14 years) "quantitative reasoning" was explored through a survey, about two thirds of them predicted that the mass of the solution will be less than the individual components (Driver & Russell, 1982). Surprisingly, students aged 14 to 17 years also demonstrated a continuous non-particulate view and were not able to conserve the mass of substances while explaining physical and chemical reactions (Ben-Zvi, Eylon, & Silberstein, 1987). Children can integrate the particulate view into their existing frameworks about perceptible properties of matter in an interesting manner. Although most secondary school students interpret chemical changes using the particle model of matter, some tend to assume that the particles possess the same macroscopic properties of that substance (Ben-Zvi et al., 1987; Driver et al., 1994). Their particulate framework thus is not consistent with the scientific model about conservation of mass. They might assume that particles shrink or melt during physical and chemical changes (Ben-Zvi et al., 1987). Without a particulate view of matter, children usually are not able to conserve the mass of substances during physical and chemical transformations of matter. This model affects the learning of the chemical principles in school (Nussbaum, 1985). If they think, for instance, that gases have no weight, they are not likely to conserve the overall mass of the gases involved in a chemical reaction (Driver, 1985).

Research suggests that developing a particulate view of matter would help students to understand, predict, and explain changes in the appearance of substances during physical and chemical changes as the reorganization of discrete particles (Driver, 1985; Nussbaum, 1985). In studying student ideas about conservation, Piaget (1973) maintained that children who initially constructed a particulate view of matter would more easily develop the idea of conservation of mass. Thus, changes in physical appearance or creation of new substances during a chemical reaction could be understood in terms of rearrangement of indestructible particles. An understanding of the particulate nature of matter is the foundation upon which the understanding of chemical and physical change rests (Driver, 1985). Therefore, in modern chemistry, the idea that matter is particulate and continuous is the fundamental model that explains any kind of transformations in matter (Driver, 1985; Nussbaum, 1985).

A study was conducted with middle school students to look at how their understanding of conservation developed while engaging with two different types of chemistry curricula. One curriculum focused on the traditional methods of instruction while the other curriculum employed concrete models present in the particulate nature of matter. The traditional chemistry curriculum focused on basic chemistry concepts, such as physical change, chemical reaction, and conservation of mass. The instruction constituted lectures, demonstrations, reading, worksheets, and questions included in the science textbook. The study also investigated the developmental trajectories of students' ideas as they participated in an inquiry-based curriculum using physical models to demonstrate a concrete view of matter as composed of particles. This curriculum also focuses on basic chemistry concepts related to physical and chemical changes and is designed to help students gain an understanding of the conservation of mass. Students are engaged in hands-on experiences, such as mixing materials to observe physical and chemical changes, and use a concrete physical model to understand and explain the changes at the molecular level. This model helps them to understand how chemical change works. The most important concept that they learn by using this physical model is that the total number of particles involved in the reaction is conserved before and after the chemical change (i.e., the total mass of the system is conserved). This concrete model is designed to help students to learn about the scientific concept of conservation of mass not as an abstract scientific principle, but as a concrete model that they could apply to explain and predict the results of experiments using the conservation of mass principle. Using this model, they can test different ratios of the reactants and predict the amount of the products in simple chemical reactions, such as the reaction between baking soda and vinegar. Students interact in small inquiry groups to conduct activities and engage in discussions around their experimental design and results.

This study was conducted in four middle-grade classrooms from two public schools in Massachusetts. Each teacher in the study taught two eighth-grade science classes and used the inquiry/model curriculum in one class and the traditional curriculum in the other. The four classes in both schools were heterogeneous in terms of ethnicity (e.g., Caucasian, African American, Latino, Asian, Haitian, mixed, etc.). A total of 73 students in four classes participated in the study, of which 38 were girls and 35 boys. The average number of students in each class was 25 at one school and 11 at the other school.

Data were collected through (a) administering a conceptual questionnaire and (b) qualitative interviews with 16 students (four from each class) before and after the curriculum. The questionnaire assesses children's understanding related to conservation of mass during physical and chemical changes (some conservation tasks are described later in the concept maps section). Semi-structured interviews lasting about 45–60 minutes were conducted to explore their ideas and emotions in-depth. Four students from each class were selected for this purpose. A variety of analytic strategies were employed to carry out a detailed analysis of the data in relation to each research question. Students' responses to the multiple-choice items in the concept questionnaire were scored. If they conserved they received a score of one, if they did not conserve the mass of substances, they were assigned a score of 0. The interview transcripts from the 16 selected students and their open-ended responses on the pre- and post-questionnaire were analyzed using qualitative strategies. Open coding and categorizing techniques (Strauss & Corbin, 1998) were used to code and organize data for the cognitive dimension in the individual responses to the open-ended questions in the concept questionnaire and interviews. Codes were grouped into themes (e.g., instances where matter is conserved) and themes were batched into categories. Field notes, analytic memos, and visual displays (Maxwell, 2005; Miles & Huberman, 1994) helped to track and compare participants' developing understanding of conservation within and across cases.

Middle school students participating in this study expressed several interesting intuitive conceptions about basic chemistry ideas. For example, most children said that air has no weight. This misconception interacted with their developing understanding of the conservation of mass in interesting and surprising ways. For instance, while predicting the mass of an open can of cola, many children thought that there would be no change in the total mass of the system because air is weightless. This misconception influenced their thinking in relation to other conservation tasks as well. While considering the mass of a sealed jar that contained a burning candle, most children thought that the total mass of the system would be less after the combustion of the candle. This prediction was grounded in the misconception that the candle wax will melt and convert into a weightless gas resulting in a decrease in the total mass of the jar. Conversely, some children thought that pressure has mass and therefore predicted that the total mass of the jar would increase due to the increased mass of the smoke after the wax turned into smoke. Another frequently expressed misconception was that mass equates density. Some students predicted a decrease in the mass of the open cola can because they thought it would lose its density after losing air or soda. Some students, on the other hand, thought that the density, and thus the mass, of the open cola can would not change as atmospheric air would enter the open cola.

Another interesting finding was that students used disparate frameworks in relation to the same conservation problem and made conflicting predictions about the mass of the system. Hence, while predicting the mass of a closed jar after the combustion of a candle, some students considered only one component of the system while making their predictions. They thought that the mass of the jar would decrease because the size of the candle would be shorter after burning. Interestingly, they used different reasoning while considering the mass of a sealed test chamber after the combustion of candles. Some students said that the total mass of the system would increase because of the smoke produced from combustion.

The findings of this study suggest that the students developed and applied localized thinking structures about conservation of mass. The scientific model of mass conservation was presented to them in the inquiry-based curriculum through the concrete particulate model and in the traditional curriculum through the abstract atomic model. Nevertheless, many students from both curriculum groups did not develop a global understanding of the conservation principle. Their preexisting skills about various phenomena, such as combustion and melting, interacted with their newly acquired ideas about the conservation of mass producing localized frameworks related to specific problems on the concept test. This means that they conserved the mass of substances on some problems while simultaneously demonstrating a non-conserving view in relation to other problems. They did not develop a coherent and generalized understanding of the conservation of matter.

The preconceptions held by these students resonated with the findings of the relevant literature (Driver, et al., 1994; Smith, 1988). The most prevalent misconceptions expressed by the students were that (a) air is weightless, (b) hot air is lighter, (c) density is equated with mass, (d) substances gain or lose mass after dissolving, (e) solids lose mass after melting, (f) the amount of melted water is greater than the amount of frozen ice, and (g) air pressure contributes to the mass of a system. Students mostly tended to focus on a single component of the system rather than considering the entire system. The existing literature on children's ideas corroborates these findings (Bar, 1986; Briggs & Holding, 1986; Carey, 1985; Driver, 1985; Driver & Russell, 1982; Smith et al., 1997; Stavy & Stachel, 1984). Some of these misconceptions persistently showed up in students' responses before and even after participating in both types of curricula.

INSTRUCTIONAL STRATEGIES FOR PROMOTING CONCEPTUAL CHANGE IN INCLUSIVE SETTINGS

Changing children's or adult's intuitive conceptions about the natural world is not an easy feat to accomplish. Different models of conceptual change have been proposed by cognitive-science experts. The conceptual change model is based on creating a dissonance between learners' existing conceptions and the new scientific ideas. According to this model, a meaningful conceptual change in children's scientific ideas requires that they must be dissatisfied with their existing ideas and appreciate the plausibility and intelligibility of the scientific conception (Hewson, 1981, 1982; Strike & Posner, 1985). Conceptual learning is generally understood as a process (Carey, 1985; Demastes, Good, & Peebles, 1996; Strike & Posner, 1992) that encompasses "conceptual assimilation" (integration of a new conception with learner's existing ideas) and "conceptual accommodation" (replacement of existing ideas by the new conception). A fundamental prerequisite for any conceptual change, nevertheless, is that students become aware of their existing ideas so that they can think about the ways in which they are using them to understand and approach any problem-solving tasks while learning science. Helping students to uncover and share their preconceptions and examine how their ideas change or develop further by making connections with other relevant ideas is thus an important goal of science instruction (Stepans, 2003).

Concept Maps and Graphic Organizers: Science teachers can employ different tools, such as concept maps, graphic organizers, visual displays, and drawings to help students to articulate and examine their preconceptions and developing scientific conceptions. These tools and strategies can be used effectively in regular as well as inclusive classroom settings as *all* children tend to carry intuitive preconceptions and need to be supported while learning scientific concepts. Nevertheless, research suggests that students with learning disabilities may have preconceptions that are "several years below those of their age peers" (Scruggs, Mastropieri, & Okolo, 2008, p. 3). Furthermore, children with mild and high-incidence disabilities also seem to face difficulty with developing critical thinking and independent reasoning associated with problem-solving tasks in science. Concept maps and graphic organizers can be particularly helpful in developing specific accommodations for students with special needs to facilitate their conceptual understanding of scientific models (Anderson-Inman, Ditson, & Ditson, 1998; McLeskey, Rosenberg, & Westling, 2010). Graphic organizers are tools to clarify and organize prior ideas so that new ideas can be integrated with relevant existing concepts (Ausubul, 1963).

Concept maps and graphic organizers have been shown to be effective in developing students' conceptual development in different content domains. Concept maps are valuable in supporting students to comprehend and represent the relationships among various concepts (Guastello, 2002; Seamen, 1990; Yates & Yates, 1990). Concept or cognitive maps serve as useful tools to visually represent the relationships among different components of a scientific model. They are mainly valuable in helping students to express their mental constructs and meaning-making in a graphic form. Connections among different concepts are displayed through lines or arrows to create a network of closely connected ideas (Novak, 1990, 1991, 1993, 1998). Different formats and designs can be used for constructing concept maps (Figure 7.1).

Concept maps can be used as assessment and pedagogic tools in science instruction because they make it possible to cover and evaluate the text more efficiently (Roberts & Joiner, 2007). Gerstner and Bogner (2009) suggest that concept maps may serve as effective tools for assessing students' learning achievement in science when used in combination with other assessment tools such as multiple-choice tests. Concept maps are also useful in assessing and displaying children's naïve preconceptions before instruction. The following concept map illustrates a common misconception among children about the mass of sugar after dissolving in water. These ideas were expressed by the middle school students in the study described above while predicting the mass of a sugar cube after it dissolved in the water.

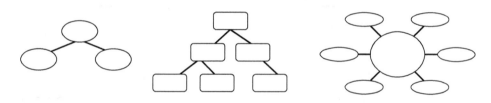

FIGURE 7.1: Concept map designs.

TASK: MASS OF A SUGAR CUBE AFTER DISSOLVING IN WATER

Sue balances a cup of water with a sugar cube outside with a cup of water with a sugar cube in it. After the sugar cube dissolves, what will happen to the pan on the right?

Preconception

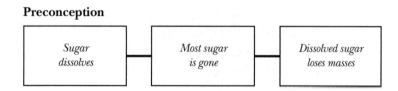

The teacher can also use a concept map as an instructional tool to exhibit the conservation of mass concept after presenting relevant hands-on activities demonstrating that the total mass of a system is conserved after sugar dissolves in water. Participation in hands-on activities and experiments is vitally important to develop students' thinking, writing, and communicating skills in science (McCarthy, 2005; Webb, 2010).

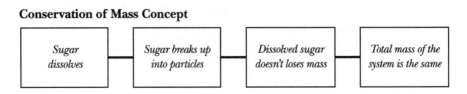

Similarly, children's misconceptions about the mass of air and the conservation of mass concepts in the context of the open cola can and combustion tasks described earlier in the discussion of the findings from the study focusing on the conversation of mass can be displayed as follows:

TASK: MASS OF AN OPEN CAN OF COLA AFTER IT GOES FLAT

Fatima balances two cans of cola, one closed and one just opened. After the open can of cola goes flat, what will happen to the pan on the right?

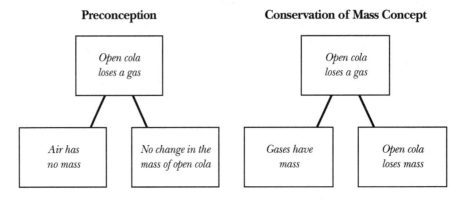

Providing opportunities to students to revisit their misconceptions about conservation of mass alongside examining the scientific model of conservation might help them to confront their alternative conceptions and see the underlying representations that contributed to their misconceptions. They can also compare the differences between their misconceptions and the scientific conceptions. Teachers in K–12 and university settings can employ concept maps to challenge and address students' intuitive conceptions and facilitate a deeper understanding of the accepted scientific models.

TASK: COMBUSTION OF A CANDLE IN A SEALED JAR

Nico balances two sealed containers, one with an unlit birthday candle and one with a lit candle. After one candle burns halfway down, what will happen to the pan on the right?

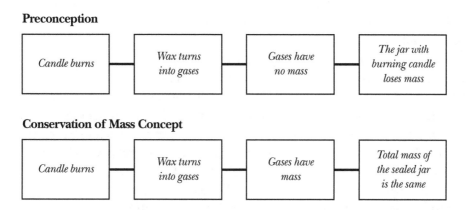

Horton, Lovitt, and Bergerud (1990) looked at the effects of graphic organizers as a pedagogical tool in a study with secondary students enrolled in content courses. Their findings suggest that graphic organizers led to higher student performance than self-study groups. Furthermore, graphic organizers were particularly useful in supporting students with learning disabilities in physical science classes (Horton et al., 1990). Similarly, Guastello (2002) reported the effectiveness of using concept maps in science instruction. Students in the intervention group used concept maps to learn and express their understanding of science concepts and performed better than the control group that was exposed to the traditional teacher-directed method. Concept mapping helped in improving low-achieving middle school students' conceptual comprehension as they actively participated in assimilating and constructing new scientific ideas.

Concept mapping has also been shown to be an effective strategy in learning, retaining, and recalling knowledge related to different domains, such as biology, chemistry, and physics (Keraro, Wachanga, & Orora, 2007; Lindstrom & Sharma, 2009; Markow & Lonning, 1998). Lindstrom and Sharma looked at the effects of using "link maps" in physics instruction in a study with first-year college students. Link maps—a physics-specific aid—were used to illustrate the interconnectedness of fundamental ideas in physics. The link maps contain fewer concepts than conventional concept maps because the purpose is to foster students' conceptual understanding of the key concepts rather than encouraging rote learning. Weekly supplementary enrichment tutorials—"map meetings"—were offered to present the materials visually using link maps. Link maps were found to be useful in advancing student understanding of abstract physics concepts and problem-solving skills, according to Lindstrom and Sharma. This study suggests that map meetings helped in creating a constructive learning environment for physics novices. Conversely, Woodward (1994) contends that generic techniques used for modifying science texts, such as direct instruction, mnemonics, and graphic organizers, may not be sufficient to deal with the "increasing complexity" of secondary science materials. Woodward argues that context-rich problem solving might be more effective in developing a rich understanding of science in secondary students with learning disabilities.

Students with learning disabilities often struggle with organizing abstract ideas and facts into coherent frameworks, and many might approach science content as a "set of isolated facts." Smith and Okolo (2010) suggest that technology-based tools could be embedded in effective research-based practices, such as "graphic organizers," "strategic and procedural support for writing," and "explicit instruction" to create better learning opportunities for students with learning disabilities. For example, Web-based writing programs provide a variety of tools to support the struggling writer, such as graphic organizers, thinking sheets, checklists, and feedback on emerging ideas; and help with organizing, revising, and editing written assignments (Smith & Okolo, 2010). Scruggs, Mastropieri, Berkeley, and Graetz (2010) conducted a comprehensive meta-analysis of the literature to assess the effectiveness of different interventions designed for special needs students. Their findings indicate that spatial and graphic organizers, mnemonic devices, and computer-assisted instruction had a significant impact on student learning in the areas of science, social studies, and English (Scruggs et al., 2010). Online tools, such as WebQuests are being increasingly used by teachers to engage children in guided learning activities. Skyler, Higging, and Boone (2007) propose that "graphic organizers, hypertext study guides, outlines, vocabulary definitions, annotated lists of Web sites, and templates for compiling information" could be used to adapt and simplify complex WebQuest tasks for students with learning disabilities.

Chin and colleagues (2010) studied the effects of online interactive concept maps—"Teachable Agents" (TA)—on students' science learning. TA is an instructional tool that engages children in teaching their agent by creating concept maps. Their agent can answer questions using the concept maps, thus enabling children to get feedback on their understanding. Students learn by teaching a computer character. The students create the concept map that is the character's "brain," and they receive feedback based on how well their computerized pupil can answer questions" (Chin et al., 2010, p. 651). The study shows that TAs enabled students to learn new concepts from their regular science lessons. Furthermore, TAs improved student understanding of causal relations among different concepts.

Concept maps can also help in vocabulary development; they can be used for creating vocabulary banks to record the scientific terms and their meanings. Students can also collect and add pictures that illustrate the concepts. Developing vocabulary webs will help them to represent how they connect different words and concepts

(Moore & Readence, 1984; Sabornie & deBettencourt, 2009). Studies suggest that vocabulary books and banks can increase reading comprehension and vocabulary recall if used as a regular pedagogical strategy (Beck & McKeown, 1991; Beck, Perfetti, & McKeown, 1982).

Strategies to Support Understanding of Scientific Texts: Comprehension of scientific texts is a persistent challenge that students face in regular as well as special education settings. According to Snow (2010), scientific language is "concise, precise, and authoritative." Sophisticated and complex scientific and technical terms can "disrupt reading comprehension and block learning" (p. 405). Science teachers need to develop and provide specific support systems to help students to understand the scientific language and become independent learners of science. Understanding the meaning of the scientific words and concepts and how they are connected to each other is essential to meaningful science learning. Many middle and secondary school students and even college students struggle with academic scientific jargon and language. Snow suggests that literacy and science education experts need to collaborate to develop strategies to assist students to "convert their word-reading skills into comprehension" while reading scientific texts. The ability to read, write, and communicate scientific ideas constitutes important features of scientific literacy (Webb, 2010). Students must be able to access the formal and specialized language of science as they engage in hands-on and minds-on science. Proficiency in scientific discourse is essential to pursuing scientific investigations. Furthermore, students must be encouraged and supported in K–12 education to ask questions and pursue inquiry-based investigations around their questions. However, providing opportunities for recording and writing about their results is equally important in science education (Webb, 2010).

Research suggests that students with disabilities lag behind their peers in middle and secondary science (Mastropieri et al., 2006). Moreover, students with disabilities struggle with scientific terms and formulae even in effective learning environments (Mastropieri et al., 2006; Scruggs, Mastropieri, Bakken, & Brigham, 1993). Differentiating science instruction by using research-based pedagogical strategies will help educators in creating different kinds and levels of supports to help *all* learners to successfully engage in science learning activities (Lawrence-Brown, 2004; Rosenberg, Westling, & McLeskey, 2008; Waldron & McLeskey, 2001).

Several studies demonstrate that strategies related to specific tasks, such as reading and concept acquisition, can facilitate student engagement and learning in specialized content domains like mathematics and science (Borkowski, Weyhing, & Carr, 1988; Hughes & Schumaker, 1991; Mastropieri, Scruggs, & Shiah, 1991). Summary skills strategy has been developed to actively engage students experiencing learning difficulties in science (Nelson, Smith, & Dodd, 1992). This learning strategy has been useful in assisting students with disabilities. Specifically, it helps students to organize, understand, and recall important ideas and information in science texts. The strategy involves identifying and summarizing the main ideas in a text in students' own words; unnecessary details are avoided so that students can focus on the key concepts presented in a scientific text. Nelson and colleagues studied the effects of summary skills strategy on elementary special education students' understanding of science text in an urban setting. They found this strategy to be highly effective in improving students' understanding as well as their writing skills.

Bakken, Mastropieri, and Scruggs (1997) examined the effects of text-structure-based reading strategies on middle school students' comprehension of science and social studies texts. Their findings suggest that this strategy is useful in helping students with disabilities who face difficulties in understanding science vocabulary, concepts, and the supporting evidence for scientific arguments. This strategy involves identifying the main ideas and their supporting evidence in a text. Students locate the main idea and supporting evidence after reading a passage and summarize this information in their own words. The study shows that text-structure-based reading strategies significantly improved students' recall of main and incidental information on immediate, delayed, and transfer assessments as compared to the traditional instructional methods in science. The authors maintain that middle school students with learning disabilities can "learn, apply, and transfer text-structure-based strategies" (Bakken et al., 1997).

The repeated-reading strategy is also beneficial for students with mild learning disabilities. Research suggests that repeated reading actively engages students in their learning and improves their reading fluency skills. This strategy involves reading a particular text multiple times. Students can also discuss the main ideas during this process. Repeated reading can serve as a useful strategy to enhance students' understanding in specialized content areas, such as science and social studies (Sabornie & deBettencourt, 2009; Sindelar, Monda, & O'Shea, 1990).

Lovitt and Horton (1994) argue for modifying content textbooks for youth with mild disabilities in inclusive classrooms. Students with learning disabilities often experience difficulties with reading and textbooks, and assimilating new information with their existing mental schemas. They also recommend that general-education teachers should be involved in the process of adapting science textbooks. Visual tools, such as graphic organizers are useful in making difficult and poorly organized passages more accessible to students. Textbooks in biology, physical and environmental science, and health are laced with sections that can be neatly organized by graphic organizers of one type or another (p. 115).

Peer-assisted learning (PAL) or peer tutoring strategies have been shown to be successful in mathematics and science learning. Mastopieri (2006) reports that peer tutoring can be employed effectively to differentiate hands-on activities to facilitate science-content learning for students with mild disabilities in inclusive middle-grade classrooms. PAL involves collaborative instruction in pair groups. Students needing academic assistance (tutees) are paired with other students who are proficient in a certain concept or skill (tutors) who use instruction and feedback related to the task to help the tutees. The tutor and tutee switch their roles over time in this process also (Calhoon & Fuchs, 2003; Sabornie & deBettencourt, 2009).

Mnemonic or memory devices have been shown to be particularly useful in different science domains like geology, biology, paleontology, and chemistry. They are highly effective in retaining and recalling information (Mastropieri, Scruggs, & Levin, 1987; Mastropieri, Scruggs, & Graetz, 2005; Sabornie & deBettencourt, 2009; Scruggs et al., 2008). For example, a useful mnemonic for remembering the names and order of the planets in the solar system is: *My Very Efficient Mother Just Served Us Nuts*. Similarly, the mnemonic for remembering the taxonomic classification scheme in biology (Kingdom, phylum, class, order, family, genus, and species) could be *King Peter can order fish, greens, salsa*.

Science educators can employ and adapt these evidence-based strategies to enhance all students' scientific literacy, conceptual development, and reasoning skills. Further research is needed to study the effects of these differentiation strategies to learn about the ways in which specific accommodations can help students with learning disabilities in science classrooms.

REFERENCES

Abraham, M., & Williamson, V. (1994). A cross-age study of the understanding of five chemistry concepts. *Journal of Research in Science Teaching, 3,* 147–165.

Anderson-Inman, L., Ditson, L. A., & Ditson, M. T. (1998). Computer-based concept mapping: Promoting meaningful learning in science for students with disabilities. *Information Technology & Disabilities, 5*(1–2).

Asghar, A., & Libarkin, J. (2010). Gravity, magnetism, and "down": College students' conceptions of gravity. *The Science Education, 19*(1), 42–55.

Asghar, A., Libarkin, J., & Crockett, C. (2001). *Invisible misconceptions: Students' understanding of ultraviolet and infrared radiation*. Presented at the GSA Annual Meeting & Exposition Program, the Geological Society of America. Boston, MA. November 5–8, 2001.

Ausubel, D. P. (1963). *The psychology of meaningful verbal learning*. New York: Grune & Stratton.

Baker, D. R., & Piburn, M. D. (1997). *Constructing science in middle and secondary school classrooms*. Boston: Allyn and Bacon.

Bakken J. P., Mastropieri, M. A., & Scruggs, T. E. (1997). Reading comprehension of expository science material and students with learning disabilities: A comparison of strategies. *The Journal of Special Education, 31*(3), 300–324.

Bar, V. (1986). *The development of the conception of evaporation*. Jerusalem: The Hebrew University of Jerusalem.

Beck, I. L., & McKeown, M. G. (1991). Conditions of vocabulary acquisition. In R. Barr, M. Kamil, P. Mosenthal, & P. D. Pearson (Eds.), *Handbook of reading research* (Vol. 2, pp. 789-814). White Plains, NY: Longman.

Beck, I. L., Perfetti, C. A., & McKeown, M. G. (1982). Effects of long-term vocabulary instruction on lexical access and reading comprehension. *Journal of Educational Psychology, 74*(4), 506-521.

Ben-Zvi, R., Eylon, B., & Silberstein, J. (1987). Students' visualization of chemical reactions. *Education in Chemistry, 24*, 117-120.

Borkowski, J. G., Weyhing, R. S., & Carr, M. (1988). Effects of attributional retraining on strategy-based reading comprehension in learning-disabled students. *Journal of Educational Psychology, 80*, 46-53.

Brewer, W. F., & Samarapungavan, A. (1991). Children's theories versus scientific theories: Differences in reasoning or differences in knowledge? In R. R. Hoffman & D. S. Palermo (Eds.), *Cognition and the symbolic processes: Applied and ecological perspectives* (Vol. 3, pp. 209-232). Hillsdale, NJ: Erlbaum.

Briggs, H., & Holding, B. (1986). *Aspects of secondary students' understanding of elementary ideas in chemistry.* Leeds, UK: University of Leeds, Center for Studies in Science and Mathematics Education.

Calhoon, M. B., & Fuchs, L. S. (2003). The effects of peer-assisted learning strategies and curriculum-based measurement on the mathematics performance of secondary students with disabilities. *Remedial and Special Education, 24*, 235-245.

Carey, S. (1985). *Conceptual change in childhood.* Cambridge, MA: Bradford Books/MIT Press.

Chin, D., Dohmen, I., Cheng, B., Oppezzo, M., Chase, C., & Schwartz, D. (2010). Preparing students for future learning with Teachable Agents. *Educational Technology Research & Development, 58*(6), 649-669.

Claxton, G. (1993). Mini theories: A preliminary model for learning science. In P. J. Black & A. M. Lucas (Eds.), *Children's informal ideas in science* (pp. 45-61). London: Routledge.

Demastes, S. S., Good, R. G., & Peebles, P. (1996). Patterns of conceptual change in evolution. *Journal of Research in Science Teaching, 33*(4), 407-431.

diSessa, A. A. (1983). Phenomenology and the evolution of intuition. In D. Gentner & A. L. Stevens (Eds.), *Mental models* (pp. 15-33). Hillsdale, NJ: Lawrence Erlbanm Associates.

diSessa, A. A. (1993). Toward an epistemology of physics. *Cognition and Instruction, 10* (2-3), 105-225.

Driver, R. (1981). Pupils' alternative frameworks in science. *European Journal of Science Education, 3*, 93-101.

Driver, R. (1985). Beyond appearances: The conservation of matter under physical and chemical transformations. In R. Driver, E. Guesne, & A. Tiberghien (Eds.), *Children's ideas in science* (pp. 145-169). Milton Keynes, UK: Open University Press.

Driver, R., Guesne, E., & Tiberghien, A. (1985). *Children's ideas in science.* Milton Keynes: Open University Press UK.

Driver, R., & Russell, T. (1982). *An investigation of the ideas of heat, temperature, and change of state of children aged between eight and fourteen years*: Leeds, UK: University of Leeds, Center for Studies in Science and Mathematics Education.

Driver, R., Squires, A., Rushworth, P., & Wood-Robinson, V. (1994). *Making sense of secondary science: Research into children's ideas*. New York: Routledge.

Gerstner, S., & Bogner, F. (2009). Concept map structure, gender and teaching methods: An investigation of students' science learning. *Educational Research, 51*(4), 425-438.

Guastello, E. F. (2002). Concept mapping effects on science content comprehension of low-achieving inner-city seventh graders. *Remedial and Special Education, 21*(6), 356-364.

Harris, P. L. (1994). Thinking by children and scientists: False analogies and neglected similarities. In L. A. Hirschfeld & S. A. Gelman (Eds.), *Mapping the mind* (pp. 294-315). Cambridge, MA: Cambridge University Press.

Hewson, P. W. (1981). A conceptual change approach to learning science. *European Journal of Science Education, 3,* 383-396.

Hewson, P. W. (1982). A case study of conceptual change in special relativity: The influence of prior knowledge in learning. *European Journal of Science Education, 4,* 61-78.

Horton, S. V., Lovitt, T. C., & Bergerud, D. (1990). The effectiveness of graphic organizers for three classifications of secondary students in content area classes. *Journal of Learning Disabilities, 23*(1), 12-22.

Hughes, D., & Schumaker, J. B. (1991). Test-taking strategy instruction for adolescents with learning disabilities. *Exceptionality, 2,* 205-221.

Keraro, F. N., Wachanga, S. W., & Orora, W. (2007). Effects of cooperative concept: Mapping teaching approach on secondary school students motivation in biology in Gucha District, Kenya. *International Journal of Science and Mathematics Education, 5,* 111-124.

Lawrence-Brown, D. (2004). Differentiated instruction: Inclusive strategies for standards-based learning that benefit the whole class. *American Secondary Education, 32*(3), 34-62.

Libarkin, J., & Asghar, A. (2002). *How children think about light and invisible light*. Paper presented at the National Association for Research in Science Teaching (NARST) Conference, New Orleans, Louisiana. April 7, 2002.

Lindstrom, C., & Sharma, M. D. (2009). Link maps and map meetings: Scaffolding student learning. *Physics Review Special Topic–Physics Education Research, 5*(1), 1-11.

Lovitt, T. C., & Horton, S. V. (1994). Strategies for adapting science textbooks for youth with learning disabilities. *Remedial and Special Education, 15*(2), 105-116.

Markow, P. G., & Lonning, R. A. (1998). Usefulness of concept maps in college chemistry laboratories: Students' perceptions and effects on achievement. *Journal of Research in Science Teaching, 35*(9), 1015-1029.

Mastropieri, M. A. (2006). Differentiated curriculum enhancement in inclusive middle school science. *The Journal of Special Education, 40*(3), 130-137.

Mastropieri, M. A., Scruggs, T. E., & Graetz, J. (2005). Cognition and learning in inclusive high school chemistry classes. In T. E. Scruggs & M. A. Mastropieri (Eds.), *Advances in learning and behavioral disabilities: Cognition and learning in diverse settings* (pp. 107–118). Oxford, UK: Elsevier.

Mastropieri, M. A., Scruggs, T. E., & Levin, J. R. (1987). Learning disabled students' memory for expository prose: Mnemonic vs. non-mnemonic pictures. *American Educational Research Journal, 24,* 505–519.

Mastropieri, M. A., Scruggs, T. E., Norland, J. J., Berkeley, S., McDuffie, K., Tornquist, E. H., & Connors, N. (2006). Differentiated curriculum enhancement in inclusive middle school science: Effects on classroom and high-stakes tests. *Journal of Special Education, 40*(3), 130–137.

Mastropieri, M. A., Scruggs, T. E., & Shiah, S. R. L. (1991). Mathematics instruction with learning disabled students: A review of research. *Learning Disabilities Research and Practice, 6,* 89–98.

Maxwell, J. (2005). *Qualitative research design: An interactive approach.* London: Sage Publications.

McCarthy, C. B. (2005). Effects of thematic-based, hands-on science teaching versus a textbook approach for students with disabilities. *Journal of Research in Science Teaching, 42,* 245–263.

McCloskey, M., & Kargon, R. (1988). The meaning and use of historical models in the study of intuitive physics. In S. Strauss (Ed.), *Ontogeny, phylogeny and historical development* (pp. 49–67). Norwood, NJ: Ablex.

McLeskey, J., Rosenberg, M. S., & Westling, D. (2010). *Inclusion: Effective practices for all students.* New York: Pearson.

Miles, M., & Huberman, M. (1994). *Qualitative data analysis. An expanded sourcebook.* London: Sage Publications.

Moore, D. W., & Readence, J. F. (1984). A quantitative and qualitative review of graphic organizer research. *Journal of Educational Research, 78,* 11–17.

Nakhleh, M. B., & Samarapungavan, A. (1999). Elementary school children's beliefs about matter. *Journal of Research in Science Teaching, 36*(7), 777–805.

Nelson, J. R., Smith, D. J., & Dodd, J. M. (1992). The effects of teaching a summary skills strategy to students identified as learning disabled on their comprehension of science text. *Education and Treatment of Children, 15*(3), 228–243.

Novak, J. D. (1990). Concept maps and Vee diagrams: Two metacognitive tools for science and mathematics education. *Instructional Science, 19,* 29–52.

Novak, J. D. (1991). Clarify with concept maps: A tool for students and teachers alike. *The Science Teacher, 58,* 45–49.

Novak, J. D. (1993). Human constructivism: A unification of psychological and epistemological phenomena in meaning making. *International Journal of Personal Construct Psychology, 6,* 167–193.

Novak, J. D. (1998). *Learning, creating, and using knowledge: Concept maps as facilitative tools in schools and corporations*. Mahwah, NJ: Lawrence Erlbaum Associates.

Nussbaum, J. (1985). The particulate nature of matter in the gaseous state. In R. Driver, E. Guesne, & A. Tiberghien (Eds.), *Children's ideas in science* (pp. 145–169). Milton Keynes, UK: Open University Press.

Osborne, R. (1980). *Force: Learning in science project*. Hamilton, New Zealand: University of Waikato.

Perkins, D. (1992). *Smart schools: From training memories to educating minds*. New York: Free Press.

Piaget, J. (1973). *The child's conception of the world*. London: Paladin.

Piaget, J., & Inhelder, B. (1974). *The child's construction of quantities: Conservation and atomism* (A. J. Pomerans, Trans.). London: Routledge and Kegan Paul.

Roberts, V., & Joiner, R. (2007). Investigating the efficacy of concept mapping with pupils with autistic spectrum disorder. *British Journal of Special Education, 34*(3), 127–135.

Rosenberg, M., Westling, D., & McLeskey, J. (2008). *Special education for today's teachers: An introduction*. Upper Saddle River, NJ: Pearson.

Sabornie, E. J., & deBettencourt, L. U. (2009). *Teaching students with mild and high-incidence disabilities at the secondary level* (3rd ed.). Upper Saddle River, NJ: Pearson Merrill Prentice Hall.

Samarapungavan, A., & Wiers, R. (1997). Children's thoughts on the origin of species: A study of explanatory coherence. *Cognitive Science, 21*, 147–177.

Schneps, M. H., & Sadler, P. (1997). *Minds of our own*. Boston, MA: Harvard Center for Astrophysics.

Scruggs, T. E., Mastropieri, M. A., Bakken, J. P., & Brigham, F. J. (1993). Reading vs. doing: The relative effectiveness of textbook-based and inquiry-oriented approaches to science education. *The Journal of Special Education, 27*, 1–15.

Scruggs, T. E., Mastropieri, M. A., Berkeley, S., & Graetz, J. E. (2010, November/December). Do special education interventions improve learning of secondary content? A meta-analysis. *Remedial and Special Education, 31*, 437–449.

Scruggs, T. E., Mastropieri, M. A., & Okolo, C. M. (2008). Science and social studies for students with disabilities. *Focus on Exceptional Children, 41*(2), 1–24.

Seaman, T. (1990). *On the high road to achievement: Cooperative concept mapping*. (ERIC Document Reproduction Service No. ED 335 140)

Sindelar, P. T., Monda, I., E., & O'Shea, I. J. (1990). Effects of repeated readings on instructional- and mastery-level readers. *Journal of Educational Research, 83*, 220–226.

Skylar, A. A., Higging, K., & Boone, R. (2007). Strategies for adapting WebQuests for students with learning disabilities. *Intervention in School & Clinic, 43*(1), 20–28.

Smith, C. (1988). *Weight, density and matter: A study of elementary children's reasoning about density with concrete materials and computer analogs*. Cambridge, MA: Harvard Graduate School of Education, Educational Technology Center.

Smith, C., Maclin, D., Grosslight, L., & Davis, H. (1997). Teaching for understanding: A study of students' preinstruction theories of matter and a comparison of the effectiveness of two approaches to teaching about matter and density. *Cognition and Instruction, 15*(3), 317-393.

Smith, S. J., & Okolo, C. (2010). Response to intervention and evidence-based practices: Where does technology fit? *Learning Disability Quarterly, 33*(4), 257-272.

Snow, C. (2010). Academic language and the challenge of reading for learning about science. *Science, 5977*(328), 393-532.

Stavy, R., & Stachel, D. (1984). *Children's ideas about solid and liquid*. Tel Aviv: Tel Aviv University, Israel Science Teaching Center.

Stepans, J. (2003). *Targeting students, science misconceptions*. Tampa, FL: Showboard.

Strauss, A., & Corbin, J. (1998). *Basics of qualitative research: Grounded theory, procedures and techniques*. Newbury Park, CA: Sage.

Strike, K. A., & Posner, G. J. (1985). A conceptual change view of learning and understanding. In L. H. West & A. L. Pines (Eds.), *Cognitive structure and conceptual change* (pp. 211-232). London: Academic Press.

Strike, K. A., & Posner, G. J. (1992). A revisionist theory of conceptual change. In R. Duschl & R. Hamilton (Eds.), *Philosophy of science, cognitive psychology, and educational theory and practice* (pp. 147-176). Albany, NY: SUNY Press.

Thagard, P. (1989). Explanatory coherence. *Behavioral and Brain Science, 12,* 435-502.

Vosniadou, S. (1994). Capturing and modeling the process of conceptual change. *Learning and Instruction, 4,* 45-69.

Vosniadou, S. (2007). The cognitive-situative divide and the problem of conceptual change. *Educational Psychologist, 42*(1), 55-66.

Vosniadou, S., & Brewer, W. F. (1992). Mental models of the Earth: A study of conceptual change in childhood. *Cognitive Psychology, 24,* 535-585.

Waldron, N., & McLeskey, J. (2001). An interview with Nancy Waldron and James McLeskey. *Intervention in School and Clinic, 36*(3), 175-182.

Webb, P. (2010). Science education and literacy: Imperatives for the developed and developing world. *Science, 5977*(328), 448-450.

Wiser, M. (1988). The differentiation of heat and temperature: History of science and novice-expert shift. In S. Strauss (Ed.), *Ontogeny, phylogeny, and historical development* (pp. 28-48). Norwood, NJ: Ablex.

Woodward, J. (1994). The role of models in secondary science instruction. *Remedial and Special Education, 15*(2), 94-104.

Yates, G. C. R., & Yates, S. (1990). Teacher effectiveness research: Towards describing user-friendly classroom instruction. *Educational Psychology, 10,* 225-238.

READING 8

SCIENCE CURRICULAR MATERIALS THROUGH THE LENS OF SOCIAL JUSTICE

RESEARCH FINDINGS

By Mary M. Atwater and Regina L. Suriel

The purpose of this chapter is to provide a practical snapshot of social justice in action with science curricula—what it looks like, why it matters, and how teachers might envision practices and orientations for their science classes. To accomplish this, the authors focus on the results and implications of studies of teachers that designed multicultural science curriculum and their results of implementing some of the activities in their science classrooms. The implications of research on multicultural science curricula have the potential for future impact on student science learning, science teacher education programs, and research. It even has more to say about the difficulty of maintaining a social justice agenda in science education.

INTRODUCTION

With the printing of Science for All Americans (SFAA) in 1990 by the American Association for the Advancement of Science (AAAS), scientists, mathematicians, engineers, physicians, philosophers, and historians called for high school graduates to be scientifically literate with the return of Haley's comet in 2061. To be scientifically literate under the SFAA ideals, students need to understand not only the tenets and approaches of science, but they also need to be literate in mathematics and technology. Then in 1993, AAAS published Benchmarks for Science Literacy as a tool for educators to use to fashion their own curricula in every state and school district. Elementary, middle school, and high school teachers, school administrators, scientists, mathematicians, engineers, historians, and learning specialists participated in the development of the benchmarks that were to be reached by students at specific grade levels, so that all students in the United States would become literate in science, mathematics, and technology. In 1996, the National Research Council published the National Science Education Standards, "a vision of science education that will make scientific literacy for all a reality in the 21st century" (p. ix). With these standards for science teaching, professional development for teachers, assessment, K–12 science contents, science education programs, and school systems, the science education community was poised to begin its work on ensuring science literacy for all students in the United States. However, the authors of this chapter do not believe that any of these documents provide philosophical and theoretical lenses to accomplish this literacy for all.

Mary M. Atwater and Regina L. Suriel, "Science Curricular Materials through the Lens of Social Justice: Research Findings," *Social Justice Pedagogy Across the Curriculum: The Practice of Freedom*, ed. Thandeka K. Chapman and Nikola Hobbel, pp. 273-282. Copyright © 2010 by Taylor & Francis Group. Reprinted with permission.

Hence, the purpose of this chapter is to provide a practical snapshot of social justice in action with science curricula that will begin to ensure science literacy for all students in the United States. Since there is very little ongoing research about science curricula, especially with an emphasis on social justice, the emphasis of this chapter will be on why social justice is important in science curricula, what social justice looks like in science curriculum materials, and how science teachers might envision social justice orientations and practices.

First, we need to define some terms such as "science literacy," "all students," "science curriculum," "critical science multiculturalism," and "social justice." People define science literacy in many ways—what a person knows, what a person can do, and what kinds of careers a person pursues. For example, SFAA (1990) delineates a scientifically literate person as

> one who is aware that science, mathematics, and technology are interdependent human enterprises with strengths and limitations; understands key concepts and principles of science; is familiar with the natural world and recognizes both its diversity and unity; and uses scientific knowledge and scientific ways of thinking for individual and social purposes. (p. xvii)

According to SFAA, a scientifically literate high school graduate (a) understands the nature of science, mathematics, and technology as a human enterprise, (b) possesses the basic knowledge about the world from the perspectives of scientists, mathematicians, and technologists, (c) understands about some of the great episodes in the history of the scientific endeavor and crosscutting themes that can serve as tools for thinking about how the world works, and (d) possesses habits of the mind. The science knowledge includes knowledge of the biophysical environment and social behavior that explains the dependency of living things on each other and the physical environment and the nature of systems, the importance of feedback and control, the cost-benefit risk relationships, and the inevitability of the side effects of technology that impact the environment and the microcultures found in the United States. Scientific habits of the mind help people from all "walks of life" (p. xiv) to deal with problems that involve evidence, quantitative considerations, logical arguments, and uncertainty. Scientifically literate people can participate with others to build and protect an open and just society.

No Child Left Behind (NCLB) legislation defines who *all* students are in U.S. school systems that garner federal funds. NCBL's purpose is to ensure "all children have a fair, equal, and significant opportunity to obtain a high-quality education and reach, at a minimum, proficiency on challenging State academic achievement standards and state academic assessments" (U.S. Department of Education, n.d., Section 1001, ¶ 1). "All," then, includes students from the following groups of students: low-achieving children in the highest-poverty schools, limited English proficient children, migratory children,

children with disabilities, Native American children, neglected or delinquent children, and young children in need of "reading assistance in the United States" (U.S. Department of Education, n.d., Section 1001, ¶ 3). The purpose of the law is to close the achievement gap between high- and low-performing children, especially the achievement gaps between White students and students of color, and between underserved children and their better served peers.

In order to meet the goals of NCLB, it is necessary to focus on the curriculum. Jackson (1992) defines curriculum as all of the educative experiences of students in schools and under the guidance of their teachers and other school personnel. Bybee and DeBoer (1994) view science curriculum as what is to be learned by students in schools. Using this definition of the science curriculum, the focus of science curriculum is then on the science knowledge to be learned, the processes or methodologies that scientists use, and the applications of science, especially the relationship between science and society and science–technology–society. However, Bybee and DeBoer's definition for science curriculum limits what questions can be raised about science curriculum issues. We would broaden the definition of science curriculum so that it includes all the educative experiences that students have related to science, mathematics, technology, and society in schools.

To broaden the questions that can be raised about science curricula, the theoretical lens of critical multiculturalism proves useful. Critical multiculturalism is a philosophical perspective that critically analyzes institutions and societal structures and examines conformity, oppression, and subjugation as the result of different cultural groups (Thomson, Wilder, & Atwater, 2001). These cultural groups are formed as a result of people's ages, social classes, disabilities, ethnicities, genders, sexual identities, geographical locations, languages, places of residences, races, and religions in the United States. Critical science multiculturalism as it relates to science education is a philosophical perspective we can use to analyze and examine the inclusion of *all* students in the learning process of science so they *all* obtain a quality science education. These analyses and examinations can help us ensure that we engage antiracist science curricula so that social justice emerges. Boyer and Baptiste, Jr. (1996) believe that a curriculum is not multicultural if social justice is not the foundational base of the curriculum. Banks (1995, 2004) defines the dimensions of multicultural education and then proposes four levels of multicultural integration into curricula: contributions approach, the additive approach, the transformation approach, and the social action approach. Level 1, the *contributions approach*, focuses on heroes, heroines, holidays, and discrete cultural fundamentals, while Level 2, the *additive approach*, includes the insertions of content, concepts, themes, and perspectives without changing the structure of the curriculum. Level 3, the *transformation approach* occurs when the very structure of the curriculum is altered so that students can view concepts, issues, events, and themes from the perspectives of many others, that is, no longer is the

Eurocentric perspective the dominant view of the curriculum. Finally, in Level 4, the *social action approach*, social justice is central in that students use their knowledge and skills to make decisions about important social issues and take action to help solve problems, including their own. Atwater (2003) delineates science examples related to Banks's dimensions of multicultural education, but until now, there have not been any research findings based on Banks's levels of multicultural integration into science curriculum materials.

According to Nieto and Bode (2008), social justice is a philosophy, an approach, and actions that embody treating all people with fairness, respect, dignity, and generosity. On a societal scale, this means affording each person the real—not simply a verbalized—opportunity to reach their potential by giving them access to the goods, services, and social and cultural capital of a society, while also affirming the culture and talent of each individual and the group or groups with which they identify. It challenges, confronts, and disrupts myths, misconceptions, untruths, stereotypes, and cultural assumptions that lead to structural inequality and discrimination based on human differences. Banks (1993) and Grant (1994) have addressed myths about multicultural education, illustrating that untruths can be based on cultural assumptions, ideas that are taken for granted to be true about members of a cultural group or the culture of individuals. Assumptions can take the form of commonplace, generally accepted statements, such as, "Middle Eastern people support terrorism," "people living in urban communities deserve to be living in these poor areas," or "teenagers should be paid minimum wage." These assumptions derive from cultural frames of reference that are based upon an individual's experiences. Most people have developed rationalizations for their cultural ideas and positions. However, most of their cultural assumptions are really beliefs with little or no evidence that they are true. As teachers, we come with cultural assumptions, and we act upon them sometimes in detrimental ways and sometimes in a manner that benefits our students and the lives of others in our communities. In science classrooms where social justice is paramount, it becomes possible for students to make changes in their lives.

For instance, science teachers with a social justice perspective are committed to providing all students with the necessary resources to learn to their full potential. Kohl (1998) describes such a teacher as "one who cares about nurturing all children and is enraged at the prospects of any of her or his students dying young, being hungry, or living meaningless and despairing lives ... . They go against the grain and work in the service of their students" (p. 285). These resources include material resources such as science books and other science curricular materials.

Equally vital are emotional resources such as a belief in students' ability and worth; care for them as individuals and learners; high expectations and rigorous demands on them; and the necessary social and cultural capital to negotiate the world. Science teachers' commitment to social justice is impacted by the school and community environments. Science teachers not only give their students resources, but draw on the talents and strengths of their students. Hence, they do not embrace a deficit perspective that has characterized much of the education of marginalized students, but shift their views to all students—not just those from privileged backgrounds—as having cultural capital that can be a foundation for their learning. It includes their languages, cultural knowledge, connections to their culture, and experiences (Bourdieu & Passeron, 1990).

UNDERSTANDING SOCIAL JUSTICE AND SCIENCE CURRICULUM ISSUES

Studies in this area have focused on either teacher beliefs or multicultural educational material alone or together using various combinations of instruments measuring beliefs and student performance (Lee, 2004; McLaughlin, Shepard, & O'Day, 1995). In a study of the relationship of science teachers' beliefs and science curricula, Lee (2004) indicated that the establishment of instructional congruence depends on successful integration of student language and culture in addition to science and literacy instruction. Furthermore, Lee found that science teachers' ability to practice instructional congruence—the integration of the nature of science with students' languages and cultures (Lee, 2003)—is a gradual process that depends in large part on teacher professional development and support. In a study that focused on the nature of teacher beliefs as determinants of teaching practices and pedagogies, Brand and Glasson (2004) showed that pre-service teachers' experiences with diversity confirmed or challenged pre-existing beliefs. The findings in the study of Latino paraprofessionals (Monzó & Rueda, 2003) expand upon and confirm Brand and Glasson's findings (2004).

When teachers attempt to integrate multicultural content into science curricula, the level of integration is well intentioned but not well informed; integration is typically at a superficial level. Using Banks's (1995) approaches to multicultural curriculum reform, Key (2000) showed that most of the participating pre-service teachers were able to integrate (a) culturally relevant examples, such as data and information, to illustrate key concepts and ideas in their academic discipline in the content integration or contributions approach and (b) content, concepts, and lessons into their curriculum without changing its structure in what Banks refers to as the additive approach. However, few of the teachers in this study transformed the curriculum to enable students to examine content from multiple ethnic and cultural perspectives (the transformation approach), and none of the teachers provided opportunities for the students to make decisions and take actions concerning civic duties (the action approach).

A critical examination of ideas about learning and curricular and instructional practices does extend its focus to pre-service teachers currently enrolled in teacher education programs; Van Hook (2002) reports pre-service teachers' attitudes toward the implementation of multicultural science curricula and suggests that their identified barriers to teaching ethnically diverse classrooms are manifestations of their negative attitudes toward these learners. Gay (1990) maintains that multicultural teachers must (a) be able to conceptualize equity as comparability or equivalence of learning opportunities for students of color instead of the same treatment for everyone, (b) be aware of their routine teaching behaviors that militate against educational equity, and (c) learn how to make regular instructional procedures more accommodating to students of color.

Using research methods to assess teachers' philosophies, we can identify science teachers' ideas about multicultural education and the effect these ideas have on science curriculum material development and the implementation of multicultural science activities in their classrooms. In the next section, you will find science activities that are examples of science teachers' attempts to incorporate multicultural education into science curriculum units and their classroom teaching.

PRACTICAL APPLICATIONS OF SOCIAL JUSTICE TO SCIENCE CURRICULA

This section of the chapter focuses on some examples of science curriculum materials created by science teachers enrolled in a Master's level science curriculum course. The following exercises are excerpted from the science lesson plans of the participants in a longitudinal qualitative study and serve as examples of their attempts to incorporate multicultural education.

Exercise I. What culturally based assumptions are being made in the following activities?

1. From an Earth Science lesson:
   Students will listen to the chorus students singing the water cycle song to the tune of "She'll Be Coming 'Round the Mountain".
   Water travels in a cycle, yes it does (yes it does).
   Water travels in a cycle, yes it does (yes it does).
   It goes up as evaporation,
   Forms clouds and condensation,
   Falls down as precipitation, yes it does (yes it does).
2. From a lesson on sound:
   The students are to listen to a classical music piece.
3. From a January 2008 N.Y. Regents Physics examination:
   While riding a chairlift, a 55-kilogram skier is raised a vertical distance of 370 meters. What is the total change in the skier's gravitational potential energy?
4. From a science lab lesson:
   For a group activity, the teacher asks:
   a. All Latino students to group together.
   b. Females to be the recorder.

Exercise II. Which of Banks's four approaches to multicultural education (Contribution, Additive, Transformative, and Action) is used in each example? If more than one approach is applicable, state why.

1. Mrs. Mary wants her students to question their image of scientists. Students are asked to draw a picture of a scientist. Mrs. Mary then conducts a discussion through a PowerPoint presentation depicting pictures of scientists from different ethnicities and cultural backgrounds.

2. In his Biology class, Mr. Harris assigns his students to construct a food web. Students can choose from any area around the world, conduct research, and then construct a food web. Students present their food web to the class.

3. As part of a graphing laboratory exercise, students are asked to read an article explaining the history of chewing gum. The article highlights ancient use by the Greeks and Native Americans. It goes on to the development and introduction of modern gum by the Mexican general Antonio Lopez de Ana to the U.S. entrepreneur Thomas Adam, Sr., who then began the mass marketing of chewing gum. Students are then asked to collect chewing time data from their lab assistants and construct a graph.

4. For a class covering the concept of the ecological footprint, students will research relevant information then take an ecological footprint quiz that surveys their habits (http://www.ecofoot.org/). After all of the students have received their scores, the class will discuss ways to improve on their own ecological footprint. Over the next five days, students will keep a journal on what they waste each day (how long they have the water running when they brush their teeth, how long it took them to take a shower, check to see if lights or computers were on and for how long, how much food did they throw away, etc.). The class will set a goal as to what the average should be for the ecological footprint. The students will then develop and distribute a newsletter containing facts on ecology, conservation, and local student change. A student club will be formed to discuss further issues with an eye toward interaction with the larger community. A club website will serve as a means of communication between the students and the community. The ultimate goal will be for the community to take action and reduce water and electricity use. This action may include the enactment of legal measures to enforce this decision.

5. In a unit on cells, Mrs. Roberts designs ten interactive, inquiry-based lessons explaining, discussing, exploring and investigating cells, cell theory and diversity, cell design and structure and ethnobotany. Lesson number eight is designed so students discover the many uses of plants by different people around the world. Students are asked to choose a plant to research and describe/explain how indigenous groups have used it for their benefit. Students are then to produce an artifact describing the plant and the culture of the indigenous group to place on the classroom bulletin board.

Exercise III. A teacher develops a unit on ecology for a students enrolled in a Life Science course. The first three inquiry-based lessons are designed to develop an understanding of the structure, function, and relationships found among ecosystems. The fourth lesson focuses on human population dynamics, characteristics, and impact. The following is an excerpt of an agenda included in the lesson design for Human Populations and Immigration.

1. Activating thinking strategy—Human population curve over the last one hundred years or a graph of the number of immigrants into the U.S.A. over the last one hundred years.
    a. Ask students discussion questions such as: What is a population? How can a population grow? What does the word immigration mean? Why is there an increase in the number of immigrants in 1910? Why is there a decline in 1930 and 1940? What new wave of immigration is currently happening?
2. Application of population dynamics/ecology to current event
    a. Students will watch carefully selected segments of a recent roundtable discussion about U. S. immigration policy, reform, and assimilation from an online base website.
    b. As students watch segments, review key components of the major bills/proposals in Congress and how they differ.
3. Congressional Town Hall Meeting—U. S. immigration policy
    a. Students placed in groups randomly through colored/numbered index cards, such as politicians, small business, big business, National Council of La Raza, undocumented immigrants, "patriotic" Americans, documented immigrants, Middle East immigrants, school officials, or teachers, and within that group they develop a stance on local immigration.
    b. Students outline group stance, develop questions for other groups and develop "rebuttal" answers to anticipated questions/points from other groups.
    c. Each group will make their case to a panel, with teacher acting as a member of Congress and moderator seeking public opinion/input on controversial issue in order to hopefully develop a comprehensive bill that can be approved by the House and Senate.
    d. Each group has 3 minutes to present their stance to the member of Congress and 1–2 minutes of questions from other groups. While each group presents, moderator must maintain a professional and respectful environment in which each group listens to stance of others, without personal attacks or disrespectful comments.
    e. At the conclusion of the meeting, students are asked to think about how the ever-growing population will affect local, national, and global environments (Prompting questions: Where will all the people live? How may this population boom impact the environment?).

f. Students will be assigned a minimal half-page reflection in which they must defend their personal stance on immigration policy and how population growth/immigration relates to ecological issues.

Questions for Reflection and Discussion

1. What are the multicultural approaches and strategies used in this lesson?
2. In what way would the lesson outcomes lead to instilling a sense of social justice in the students?
3. What is the role of multicultural education in the promotion of social justice?

Note: suggested answers to Exercise II—Banks's multicultural curriculum approaches

1. Contribution
2. Additive
3. Contribution
4. Action
5. Additive-Transformative

Points of Inquiry

- Read the documents "Science for All Americans," "Benchmarks for Science Literacy," and the "National Science Education Standards." What kind of students do these documents imagine? What kinds of knowledge, skills, and dispositions do they privilege? Atwater and Suriel question the documents' philosophies and theories—what do you think?

- Often, people talk about scientific knowledge as neutral and logical: Free from political constraints and social issues. What kinds of historical and sociocultural evidence can you find to problematize this approach to scientific knowledge?

- Review Banks's levels of multicultural education and compare them to the science program in the school you work: What different levels are in play? Why?

- How could you and your colleagues create a professional development environment that supports Lee's idea of instructional congruence?

Points of Praxis

- What kind of curriculum could you design that both meets the standards and places scientific knowledge in a sociocultural and social justice context?

- What activities could your students engage in that would allow them to see that scientific knowledge has been used simultaneously as a language of power, giving access to marginalized people, and also a tool of oppression that has been used to marginalize people?

- Rewrite definitions of science literacy to reflect a social justice orientation.
- What knowledge do your students already have? Design a series of activities to elicit their family, community, and intellectual funds of knowledge.

REFERENCES

American Association for the Advancement of Science (AAAS). (1990). *Science for All Americans.* New York: Oxford University Press.

Atwater, M. M. (2003). Race and gender in science education: A reconceptualization. In R. Revere (Ed.). *Gender, race and science: Transforming the curriculum* (pp. 12–23). Baltimore, MD: Institute for Teaching and Research on Women.

Banks, J. A. (1993). Multicultural education: Development, dimensions, and challenges. *Phi Delta Kappan*, *75*(1), 22–28.

Banks, J. A. (1995). Multicultural education: Historical development, dimensions, and practice. In J. A. Banks (Ed.). *Handbook of research on multicultural education* (pp. 3–24). New York: Macmillan.

Banks, J. A. (2004). Multicultural education: Historical development, dimensions, and practice. In J. A. Banks, & C. A. M. Banks (Eds.), *Handbook of research on multicultural education* (2nd ed.). (pp. 3–29). San Francisco: Jossey-Bass.

Bourdieu, P., & Passeron, J. C. (1990). *Reproduction in education, society, and culture.* (R. Nice, Trans.). Thousand Oakes, CA: Sage.

Boyer, J. B., & H. P. Baptiste, Jr. (1996). *Transforming the curriculum for multicultural understandings: A practitioner's handbook.* San Francisco: Caddo Gap Press.

Brand, B. R., & Glasson, G. E. (2004). Crossing cultural borders into science teaching: Early life experiences, racial and ethnic identities, and beliefs about diversity. *Journal of Research in Science Teaching, 41*(2), 119–141.

Bybee, R. W., & DeBoer, G. E. (1994). Research on goals for the science curriculum. In D. L. Gabel (Ed.), *Handbook of research on science teaching and learning* (pp. 357–387). New York: Macmillan.

Gay, G. (1990). Teacher preparation for equity. In H. P. Baptiste, Jr., H. C. Waxman, J. W. de Felix, & J. E. Anderson (Eds.), *Leadership, equity, and school effectiveness* (pp. 224–243). Newbury Park, CA: Sage.

Grant, C. A. (1994). Challenging the myths about multicultural education. *Multicultural Education*, *2*(2), 4–9.

Jackson, P. W. (1992). Conceptions of curriculum and curriculum specialists. In P. W. Jackson (Ed.), *Handbook of research on curriculum* (pp. 3–40). New York: Macmillan.

Key, S. G. (2000). Applications of "multiculturalism" demonstrated by elementary preservice science teachers. Paper presented at the Annual Meeting of the American Educational Research Association, April, New Orleans, LA.

Kohl, J. (1998). Afterword: Some reflections on teaching for social justice. In W. Ayers, J. A. Hunt, & T. Quinn (Eds.), *Teaching for social justice* (pp. 285-287). New York: Free Press and Teachers College Press.

Lee, O. (2003). Equity for culturally and linguistically diverse students in science education: Recommendations for a research agenda. *Teachers College Record, 105*(3), 465-489.

Lee, O. (2004). Teacher change in beliefs and practices in science and literacy instruction with English language learners. *Journal of Research in Science Teaching, 41*(1), 65-93.

McLaughlin, M. W., Shepard, L. A., & O'Day, J. (1995). Improving education through standard-based reform: A report by the National Academy of Education Panel on Standard-based Education Reform. Stanford, CA: The Academy.

Monzó, L. D., & Rueda, R. (2003). Shaping education through diverse funds of knowledge: A look at one Latina paraeducator's lived experiences, beliefs, and teaching practice. *Anthropology & Education Quarterly, 34*(1), 72-95.

National Research Council. (1996). *National science education standards*. Washington, DC: National Academy Press.

Nieto, S., & Bode, P. (2008). *Affirming diversity: The sociopolitical context of multicultural education* (5th ed.). Boston: Allyn & Bacon.

Thomson, N., Wilder, M., & Atwater, M. M. (2001). Critical multiculturalism and secondary teacher education programs. In D. Lavoie (Ed.), *Models for science teacher preparation: Bridging the gap between research and practice* (pp. 195-211). New York: Kluwer.

U.S. Department of Education. (n.d.) *Title I: Improving the Academic Achievement of the Disadvantaged*. (Retrieved April 18, 2008, from: http://www.ed.gov/policy/elsec/leg/esea02/pg1.html)

Van Hook, C. W. (2002). Preservice teachers' perceived barriers to the implementation of a multicultural curriculum. *Journal of Instructional Psychology, 29*(4), 254-264.

READING 9

THE BRAINS-ON APPROACH TO SCIENCE

By Ann Finkelstein

> 聞而忘 What I hear I forget.
> 視而記 What I see I remember.
> 行而知 What I do I understand.
> —Chinese proverb

SCIENCE IS NOT JUST FOR NERDS

Science is often perceived as a boring subject, although I cannot understand why. Science is the study of life and death, the oceans and the earth, the flora and the fauna. Science investigates the smallest particles of matter and the breadth of the universe. Science explains why Michael Jordan can jump so high, and just how difficult it is to hit a knuckle ball. Scientific research has developed technologies to allow astronauts to survive in outer space. Science holds the answers to why our children may look like us, or why they may not. The study of science has yielded the cure or prevention for horrible diseases. Science can help us determine ways to keep this planet a safe and beautiful place for many generations to come. What is boring about that?

Science is an extremely creative subject, and yet the creative aspects of science are often overlooked. Many people apply the word "creative" to art, music, dance, poetry, and fiction, rather than to science. Finding the solution to any problem, whether it is how to paint a landscape or how to map the path of a comet requires creative thought. Creative problem solving requires two steps. First, the solution to the problem must be imagined. The landscape must be envisioned by the artist, or the elliptical orbit of the comet must be hypothesized by the astronomer. The second creative process is imagining unique ways to solve the problem. The artist finds a new and different way to portray the scene, while the astronomer uses the comet's previous locations to predict how it moves through space. Scientific thought is organized by formal problem-solving techniques. Logical thinking no more limits the creative aspects of science than the methods for applying paint to canvas limit the creativity of artists. Without creative thought, there would be few scientific advances.

Science is on the brink of countless amazing and wonderful discoveries, and the media has sensationalized some recent breakthroughs. The famous sheep, Dolly, was "cloned" by fusing a mammary cell to an unfertilized, DNA-free egg.[1] Dolly's existence caused

Ann Finkelstein, "The Brains-On Approach to Science," *Science is Golden: A Problem-Solving Approach to Doing Science with Children*, pp. 3-12. Copyright © 2002 by Michigan State University Press. Reprinted with permission.

public outcry and discussions of the moral implications of this new technology. While several kinds of animals have been cloned successfully, human cloning remains technically difficult, impractical, and less efficient than traditional methods. Popular fiction often portrays scientists as the bad guys planning nefarious plots with evil purposes. Let me try to reassure you here. Scientific research is very carefully regulated. The use of chemicals, radioisotopes, animals, recombinant DNA technology, and human tissues is closely monitored. Laboratories that fail to comply with these regulations are not allowed to continue operating. We tend to fear things we do not understand, but science is not incomprehensible, and it need not be feared. Showing children how to plan experiments and analyze data will prepare them to understand and scrutinize future scientific developments. Establishing a positive attitude about science allows children to examine technological advances logically without anxiety coloring their thoughts.

Science *is* a difficult subject. Professional scientists investigate some of the most interesting, challenging, and important problems that the world has ever known. Science, however, can be done at many levels. Children can learn to appreciate the beauty of analytical thought, the perfection of nature, and the thrill of investigation without being intimidated by complicated details. There is plenty of time for them to learn about the Higg's boson, should they so desire. Now is the time to communicate that science is fun, understandable, useful, and interesting. You don't have to be a rocket scientist to do experiments. All you need is an open mind and the desire to solve problems. Nearly all of the references I have used in the preparation of this book can be found in the children's section of the public library. Research will undoubtedly be required, but it isn't necessary to wade through a graduate-level physics book to help design an experiment to answer "How does a kite fly?"

Science is based on logical arguments and simple strate-

While several complex scientific concepts are involved in flying a kite, the main reason is simple. *The wind pushes it up*. The pressure of the wind against the inclined surface of the kite generates lift. We have all experienced Newton's third law of motion: "To every action, there is an equal and opposite reaction." In this case, the action of the wind when it is deflected downward off the flat surface of the kite causes a reaction of the kite getting pushed up. The angle of the kite against the wind is important. The kite strings maintain the appropriate angle.[2] Think of all the experiments that could be done investigating the shape, size, weight, construction materials, and steering capabilities of a kite.

Reading 9—The Brains-On Approach to Science | 185

gies. The laws of nature govern everything from the smallest subatomic particles to the movement of the planets. There are relatively few scientific principles, and most of them can be understood in a way that is simple and intuitive. The key to much of biology is the survival of the fittest, and the need for animals to develop a niche where they can live, eat, and reproduce. Perhaps more natural laws govern the physical world, but there are many interesting systems that can be presented in a user-friendly way. Two of the questions I collected from a fifth grade student illustrate this concept: "Why are baby animals so cute?" and "How does a musical instrument make sound?"

First, how do Darwin's theories apply to baby animals? Cuteness has survival benefits. Soft fur or downy feathers keep small animals or birds warm. The spots on a fawn act as camouflage when the baby hides from predators, and the long legs on a zebra foal enable it to run with the herd shortly after birth. When young animals play, they are developing important survival skills such as running or hunting.[3] Of course, the details of a biological system may have to be researched, but the underlying concept of the need to survive puts many details into perspective.

Similarly, young children can design experiments to investigate how a musical instrument makes sound by examining the ideas of vibration and sound waves. When we hear sound, we perceive vibrations. These vibrations move through the air as sound waves, which are detected by our ears. Slower vibrations make lower sounds; faster vibrations make higher pitches.[4] This is fertile ground for designing experiments. Noise-making vibrations can be produced by twanging stretched rubber bands, tapping bottles filled with different amounts of water, rubbing a moistened finger around the rim of a stemmed glass, ringing bells, etc.

Science is further simplified by the use of controlled experiments. If an experiment is properly planned, it should yield data that are easy to analyze. The results of a controlled experiment should point to the correct answer. Here is an example.

The children in a first-grade soccer league wear reversible blue and gold T-shirts. Each team is assigned a color for each game, and players on opposing teams are differentiated by the color of their shirts. The teams switch colors from week to week. After the first three games, one boy wondered, "Why does the blue team always win?" Adults realize that these results are simple coincidence. Eliminating misleading chance occurrences in scientific experiments is important, however. The best way to avoid being deceived is to do controlled experiments. How can the "blue team theory" be tested? The simplest experiment would involve only two teams. The teams should play soccer six or eight times. (The large number of games helps "average out" day-to-day variations in the level of play.) As all players will probably improve their soccer skills over the course of the experiment, each team should wear the coveted blue T-shirt for alternate games. I suspect that by the end of the experiment, there would be no correlation between winning and the color of the jersey. If all the players believed that the blue shirt conferred good luck, it is possible that the added self-confidence would affect the outcome of the game. The best experiment would involve players who did not know that the "blue team theory" was being tested. (Note added in proof: the gold team won the fourth and fifth games.) If the experiment is bounded by controls, it is easier to draw correct conclusions from the data. (Designing controlled experiments is explained more completely in Chapter 4.)

Young children can be introduced to science in a variety of ways. In this book, I propose a method that starts with children's questions about science. Students then design experiments to answer their questions, and learn how to analyze and present their data. I favor this technique because the "Brains-On Method" is a complete representation of the scientific method, and because students are likely to feel responsible for and interested in experiments of their own creation. The brains-on method is just one approach to teaching science, but the techniques described in this book can also be used to enhance and clarify other methods. All students should be encouraged to ask questions. Experimental controls can be added to science demonstrations. The suggestions for graphing, preparing laboratory notebooks, and creating posters can be applied to any experiment. Most importantly, children should have a personal stake in the experiment. Children are naturally curious, and science is a way they can learn about their world. To use the words of Dr. Bruce Alberts, president of the National Academy of Sciences, "learning science is something that students do, not something that is done to them."

THE BRAINS-ON APPROACH HELPS STUDENTS CREATE THEIR OWN EXPERIMENTS

The brains-on method goes one step beyond hands-on science activities for children. Not only do the students do the experiment, they first conceive of the idea, refine their idea, plan the experiment, perform a controlled experiment, and analyze and present the results. This is easier than it seems. It does not require trained professionals, and you can attempt it in your own home or classroom. Here is a brief overview of the brains-on approach to science.

1. **Start with science questions asked by children**

The world around, above, below, and within us is fascinating, and children are naturally curious about themselves and their environment. I tried to tap into that curiosity by collecting questions from elementary school students. As a scientist, I was delighted with the quality of the questions and gratified to know the answers to some of them. The questions varied in subject, scope, and difficulty. Some were easily answered. "How many grams are in a pound?" (There are about 454 g/lb.) Some questions hint at problems of such importance that their answer may someday be worthy of a Nobel Prize. "How does medicine know where to go?" (The efforts of many talented scientists and many research dollars are currently being directed at "magic bullet" therapies in which anti-cancer drugs are targeted directly at the cancerous tissue.) By starting with questions asked *by* students, we start with their interest, and show them how to follow up on it.

2. **Show how students can turn questions into experiments**

Chapter 2 is about questions. Several examples are given of how questions can be clarified and refined so that they can serve as the nucleus of an experiment. Of course, not every question leads to an experiment, but if enough questions are collected, there are sure to be some that are suitable for experimentation. How to turn a clear and well-thought out question into an experiment is discussed in Chapter 3. With some adult guidance, a group of children can work together to build a model of how they think their experimental system may work. They propose possible answers to their question, and think of ways to test which of their ideas are correct.

3. **Plan the experiment**

In Chapter 4, I explain how to do a controlled experiment. Experimental controls are often overlooked in elementary science, but controls make the results more meaningful. Controls frame the logical argument. Negative controls measure background, and positive controls test if the experiment is working. Many examples of negative and positive controls are presented as well as reasons for doing them.

4. **Do the experiment**

Doing the experiment and collecting the data are two of the most fun parts of the experimental process. Data must be collected in an organized and careful manner. Chapter 5 covers doing the experiment, collecting the data, and keeping a laboratory notebook.

5. **Figure out what was learned**

How to make sense out of the data is discussed in Chapter 6. Lots of numbers can be baffling. Non-numeric data can be confusing too. Making sense of the experimental results and understanding the data is fun and rewarding. Topics include graphing, experimental error, qualitative analysis, and presentation of data. These subjects are discussed in a way that is appropriate for experiments done by young children.

6. **Ask more questions**

Did the results of the experiment lead to more questions? Is it possible to follow up on any of these questions? How could the experiment be improved? Was something unexpected learned from the experiment?

The Brains-on Approach

The brains-on method is diagrammed in Figure 9.1. Some people call this the scientific method. The ideas presented in the flow chart are simple. Everybody has problems. We all ask questions. We have done library research or looked up answers to questions. The terminology used in the problem-solving path may be unfamiliar, but we all use problem-solving techniques. Each step is described in detail in later chapters of this book.

Asking a question is the first step in problem solving because the question defines the problem. A question can be analyzed to determine if the answer will be found more readily by experimentation or by researching the findings of others. In some cases the question must be refined, the terms must be defined and clarified before it becomes a testable question. A testable question suggests a hypothesis or model. A hypothesis can be tested by experimentation. If the experiment is properly controlled, it will yield data that should, upon analysis, answer the question. Some questions can be answered directly (see heavy dashed line). Sometimes the answer will lead to further questions or problems (light dashed line). This book will provide a point-by-point description of how to implement the brains-on method starting with questions asked by elementary school students.

PROBLEM SOLVING IS NOT AS DAUNTING AS IT SEEMS

People solve problems using this method every day. The process is natural to humans. We may not think about solving problems in a formal, diagrammatic manner. We just do it. Children use problem-solving techniques at an early age. For example, one child covets a toy another child is using. The two children can try a number of strategies to get and keep the toy, and probably only some of these techniques will be acceptable to their parents. The problem is getting and keeping the toy. The first question is "How can I get the toy?" A more refined question might be "Can I get the toy by grabbing it?" or "Can I get the toy by asking for it?" The child will brainstorm a number of methods for obtaining the toy. The attempt to commandeer the

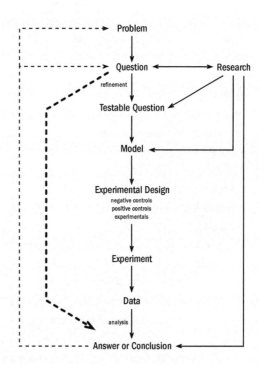

FIGURE 9.1: The Problem-Solving Flow Chart.

toy is the experiment. The data or results are whether or not the attempt is successful. Both children will imagine additional experiments.

Although problem solving is a natural ability of humans, this skill can be refined. People who are good at solving problems tend to achieve success. The best part of developing problem-solving skills is that these techniques can be applied to many situations.

When I was in graduate school, my advisor told me that the methods I was using in the laboratory would be out-of-date by the time I had finished writing my thesis. He said learning how to solve problems was most important because I could use that skill for the rest of my life. He told me how one of his former graduate students decided to manage his family's farm after completing his doctorate. While this may not be the usual path for someone trained in the academic sciences, the problem-solving skills this former student learned in graduate school would help him in whatever career he chose.

Because problem solving is inherent to human nature, young children can be encouraged to develop these skills at an early age. In this book, problem-solving skills are applied to scientific questions. I encourage parents and teachers to create an environment for doing science that is interesting and challenging, so children can apply their creative talents to learning about the world around them.

What are the Advantages to a Problem-solving Approach?

Many methods can be effectively used to teach children science. Why should parents and teachers expend the extra effort required to develop a problem-solving approach to education? Students are interested in the project because people are naturally interested in their own ideas. The students invent the science for themselves. They start with their own questions and figure out, with help, how to turn those questions into experiments. The students determine which scientific concepts they are studying. They analyze their data and draw conclusions for themselves. The experiment becomes the students' creation, not something imposed by an external source.

Problem solving is empowering for students because it encourages logical and critical thinking. The process of refining a question that I describe in Chapter 2 requires that students analyze their ideas and use specific language. Students can apply problem-solving techniques to all aspects of education and life. The ability to analyze data and draw one's own conclusions has value far beyond the completion of an elementary school science project. The problem-solving approach encourages creativity because the students have to figure out how to answer their own questions.

The brains-on method described in this book avoids the performance aspect of demonstrating experiments to students. If the experiment does not "work," it is not necessarily the demonstrator's fault. Everyone has worked together to plan and do the experiment; everyone can work together to fix it. In the real scientific world, not every experiment "works," but unsuccessful experiments may have value. Experiments can usually be repeated and improved. Scientific mistakes and failures sometimes lead to important discoveries. Albert Einstein said, "Anyone who has never made a mistake has never tried anything new."

In the following chapters of this book, I provide a step-by-step description of how to implement the brains-on approach. Most of the examples were derived from questions asked by elementary school children. The science questions I collected from elementary school children are listed in Appendix 1. Appendix 2 is a sample laboratory notebook.

NOTES

1. I. Wilmut, A. E. Schnieke, J. McWhir, A. J. Kind, and K. H. S. Campbell, "Viable Offspring Derived from Fetal and Adult Mammalian Cells," *Nature* 385 (1997): 810–13; Alexander Newman, "Double Takes," *National Geographic World* 289 (1999): 12–15.
2. Maxwell Eden, *Kiteworks: Explorations in Kite Building and Flying* (New York: Sterling Publishing Company, Inc., 1989); Jack Challoner, *Make It Work! /Flight* (New York: Scholastic, 1995).
3. John Bonnett Wexo, "Animal Babies," *Zoobooks* 13 (1996): 1–17; John Bonnett Wexo, "Baby Animals 2," *Zoobooks* 11 (1994): 1–17.
4. Joanna Cole, *The Magic School Bus in the Haunted Museum: A Book about Sound* (New York: Scholastic, 1995); Neil Ardley, *The Science Book of Sound* (San Diego: Harcourt Brace Jovanovich, Publishers, 1991); Andrew Haslam, *Make It Work! Sound* (Chicago: World Book, 1997).

SECTION III — QUESTIONS TO PONDER ...

1. What is STS, and why is it important?

2. Why should I differentiate my science lessons?

3. Why do I need to consider social justice when teaching science?

4. What social justice factors do I need to consider when choosing and using science curricula?

Section IV

SCIENCE EDUCATION AND SCIENCE LEARNERS

Objectives

- Differentiate between brains-on and hands-on.
- Synthesize factors important in teaching English language learners (ELL) science.
- Identify why it is important to consider the needs of rural students when teaching science.
- Differentiate teaching science in an urban setting versus a rural setting.

Key Concepts

brains-on: The opposite of hands-on activities. Usually involves abstract concepts and ideas that cannot be manipulated in a physical manner.

English-language learner: ELL is a broad term used to describe any learner who is studying to learn the English language. Unlike bilingual education or a dual-language education where the instruction is in two languages, students in an ELL class might be native speakers of many different languages with the common factor for instruction being that all the students are learning English.

Introduction

This section of the book addresses ideas important to individual learners. The idea of balancing hands-on actvities with brains-on activities to stimulate, engage, and develop critical thinking skills.

The diverse and changing demograpics of the learner dictate that we must start addressing the needs of ELL students as well as students in rural areas. How do you foster learning when you are a monolingual speaker teaching science to a class of ELL students? This is important to consider as our classrooms are consisting more and more of students with diverse languages.

When all the science instruction is over, how do you know how much the learner has learned or how effective you were as a teacher? Assessment is key to gauging how much the learner has gained and how effective the instructional strategies were.

READING 10

FOSTERING SCIENTIFIC REASONING AS A STRATEGY TO SUPPORT SCIENCE LEARNING FOR ENGLISH LANGUAGE LEARNERS

By Cory Buxton and Okhee Lee

ABSTRACT

We present the conceptual and analytical frameworks for considering the development of scientific reasoning with English language learners (ELLs). We provide examples of the implementation of these frameworks using data on the reasoning complexity of 73 third-grade and 81 4th-grade ELL students as they participated in individual, task-based, interviews on the topics of measurement and transfer of energy in the contexts of school activities and home connections. Our framework builds upon two distinct research traditions: (1) experimental research in developmental and cognitive psychology and (2) research on diverse student groups in science education. Analysis pointed to several consistent patterns in students' reasoning across demographic subgroups of English for speakers of other languages (ESOL) level, home language, and gender. We also highlight the methodological challenge of considering student reasoning in both home and school contexts.

INTRODUCTION

The challenge of ensuring that all students achieve high academic standards in science has become increasingly complex as the U.S. student population becomes more culturally and linguistically diverse while simultaneously facing increasing demands of high-stakes testing and accountability under the *No Child Left Behind* Act. Teachers of ELL students face the especially challenging task of aiding these students as they struggle to simultaneously develop academic content knowledge and English language and literacy skills (August & Hakuta, 1997; Wong-Fillmore & Snow, 2002).

Reform documents addressing the academic needs of ELL students point to three language learning goals that must be prominent to support school success: (1) to use English to communicate in social settings, (2) to use English to achieve academically in all content areas, and (3) to use English in socially and culturally appropriate ways (Teachers of English to Speakers of Other Languages [TESOL], 1997, 2006). It is the second goal, learning to use English to achieve academically in the content areas, which frequently becomes the greatest struggle for ELL students and their teachers. To succeed academically, ELL students need to develop English language and literacy skills through content area instruction at the same time they are mastering the content goals in an emergent language. Failure to accomplish both of these tasks in unison usually gives rise to a growing achievement gap when ELL students are compared with their fluent English-speaking grade-level peers.

Cory Buxton and Okhee Lee, "Fostering Scientific Reasoning as a Strategy to Support Science Learning for English Language Learners," *Teaching Science with Hispanic ELLs in K-16 Classrooms*, ed. Dennis W. Sunal, Cynthia Szymanski Sunal, and Emmett L. Wright, pp. 11-36. Copyright © 2010 by Information Age Publishing. Reprinted with permission.

Content area instruction, such as science instruction, has the potential to provide a robust context for meeting learning goals in both academic content and English language and literacy development. For ELL students, however, content area instruction often takes a back seat to language instruction, rather than being viewed as an integral part of language and literacy development. Many teachers continue to assume that ELL students must first master English before learning academic content (Cochran-Smith, 1995). As additional challenges, teachers in the elementary grades are frequently not prepared to teach science content at the level of sophistication required in current state science standards, nor are they sufficiently prepared to meet the learning needs of ELL students (National Center for Education Statistics, 1999). This combination of teacher beliefs, practices, and preparation almost inevitably leads ELL students to fall behind their English-speaking peers in content area learning (August & Hakuta, 1997).

In addition to these instructional challenges, teachers of science face curricular challenges as they are asked to conceptualize and present students with a coherent picture of an ever-growing list of content topics across a broad range of science disciplines. Current science standards documents have been critiqued for lacking a compelling rationale for what content should be taught (Eisenhart, Finkel, & Marion, 1996), and for fostering a broad coverage model that fails to differentiate standards according to their cognitive complexity (diSessa & Sherin, 1998). More recently, the National Research Council synthesis on learning in the context of school science, *Taking Science to School* (NRC, 2007), has argued that the standards fail to present the core ideas in each discipline in ways that would allow students to engage in sustained investigations at increasingly sophisticated levels of complexity over time.

The current U.S. curricular model, often described as "a mile wide and an inch deep," is detrimental to all students, but can be particularly problematic for ELL students. Students are not given adequate time to explore concepts in depth or from multiple perspectives in ways that allow the practice and application of new academic language skills. All too quickly, instruction has moved on to the next topic, often leaving ELL students floundering both conceptually and linguistically.

In the current climate of assessment and accountability, another significant barrier to supporting the simultaneous development of science content knowledge and academic language is that teachers have tended to become overly focused on the content standards they need to teach, while failing to give adequate attention to how their students are making sense of what is being taught. As is the case for instructional and curricular practices, these assessment practices may be detrimental to all students, but are particularly limiting for ELL students. For example, on the 2005 National Assessment of Educational Progress (NAEP) science results, only 28% of 4th-grade ELL students achieved at the basic level or higher, with only 4% scoring at the proficient level, as compared to 71% of 4th-grade non-ELL students achieving at the basic level or higher, with 31% scoring at the proficient level. Greater awareness of what students are thinking and saying, and particularly an awareness of how students are learning to think and speak scientifically, can help the teacher focus on both the challenges and the progress of ELL students as they work to simultaneously develop science content knowledge and academic language. Additionally, by attending to how students express themselves during science tasks, such as the examples that students use and the questions they ask, teachers can develop a better understanding of the linguistic and cultural resources that ELL students bring to the science classroom, as well as the difficulties students may continue to have with science concepts and scientific language usage even after instruction.

CONCEPTUAL FRAMEWORK: MULTIPLE PERSPECTIVES ON STUDENT REASONING

As we considered strategies for how to work with teachers of ELL students on the complex curricular, instructional, and assessment issues that we have outlined above, we continued to return to the need for teachers to more closely attend to the question of what their students were thinking and the prior knowledge and experiences upon which they were basing those thoughts. This led us to begin conceptualizing a framework of student reasoning grounded in both the learning sciences and research on home language and culture in diverse communities. We realized that the questions we wished to ask about science content and academic language learning of culturally and linguistically diverse students needed to draw upon multiple research traditions. We took as our focus the development of students' scientific reasoning, which we interpreted to mean the logical thought processes used for problem solving during science tasks (Lehrer & Schauble, 2005). We arranged scientific reasoning along a continuum of complexity, ranging from the generation and elaboration of simple statements and assertions at the lower end, to providing causal justifications and explanations of underlying mechanisms at the higher end. We settled upon a framework for analyzing and interpreting the reasoning of ELL students by leveraging the lessons learned from two distinct scholarly traditions: (1) experimental research in developmental and cognitive psychology and (2) research on diverse student groups in science education. Historically, these traditions have largely worked in isolation from one another. If we are to more fully understand the scientific reasoning of culturally and linguistically diverse students, we need to build upon the lessons that have been learned across seemingly disparate paradigms.

Developmental and Cognitive Psychology Perspective on Reasoning

Research on reasoning has its foundations in the highly controlled experimental studies of classical developmental and cognitive psychology. Building on the work of Piaget (Inhelder & Piaget, 1964; Piaget, 1973), reasoning has been construed as the required thinking skills for doing science inquiry through experimentation (Keil & Wilson, 2000), evaluating evidence (Klahr, 2000), and engaging in argumentation in the service of promoting scientific understanding (Kuhn, 1991). From this perspective, reasoning is viewed as an individual cognitive process.

Historically, there have been two main approaches used to study and interpret reasoning, sometimes referred to as "experimentation strategy" and "conceptual change" (Schauble, 1996). The first approach, experimentation strategy, has used a domain-general focus on reasoning and problem solving strategies applicable to a range of science and everyday tasks (Kuhn, 1990; Metz, 1991). The focus has largely been on studying the strategies that individuals develop for generating and interpreting evidence. These strategies include hypothesis generation, control of variables, and evaluation of evidence. This research has used *knowledge-lean tasks* designed to isolate general skills and strategies in the absence of required domain-specific knowledge. The second approach, conceptual change, has focused on a more domain-specific approach to the development of conceptual knowledge within particular science topics (Carey, 1987; Keil & Wilson, 2000; Smith, Maclin, Houghton, & Hennessey, 2000). This research has used *knowledge-rich tasks* designed to examine the content and structure of students' domain-specific reasoning and problem solving. Recently, some of the researchers within this tradition have shifted their attention from laboratory to classroom settings in order to examine how reasoning develops when students are provided with effective instruction over a sustained period of instructional time in a less controlled school context (Lehrer & Schauble, 2005; Metz, 2004; Toth, Klahr, & Chen, 2000).

Research from developmental and cognitive psychology provides several important lessons for examining the reasoning of ELL students. First, this research points to the challenge for students (and their teachers) of substantially re-crafting their prior ideas and conceptions, and the corresponding importance of identifying students' prior knowledge on science topics (diSessa & Sherin, 1998). Accessing the roots of prior knowledge in meaningful ways must necessarily lead to connections to students' home languages and cultures. Failure to consider the role of home language and culture in prior knowledge formation ignores or even negates critical tools that ELL students have available to construct new understandings of the world.

Second, this research has shown that a focus on generic process skills (e.g., observation, description, prediction, inference) is generally a weak way to promote student reasoning as compared to knowledge-rich activities within a particular domain (NRC, 2007; Zimmerman, 2000). Despite these findings, knowledge-lean science learning activities focusing on the processes of inquiry remain very common in school science instruction at all levels. ELL students are particularly likely to receive knowledge-lean science instruction, due to the misguided belief that ELL students need to be sheltered from more rigorous content knowledge until their English language skills improve. Instead, engaging in complex science tasks is essential for the development of science content knowledge, scientific reasoning skills, and academic English.

Still, such a paucity of research from the developmental and cognitive psychology tradition has been conducted in typical school contexts, let alone with large numbers of ELL students, that many questions remain about how the reasoning of linguistically and culturally diverse students develops in school learning environments. While this tradition brings clarity through highly controlled experimental design, our framework must also account for the unique needs of ELL students in less controlled contexts, such as elementary school learning environments and the students' home and play environments.

Diversity Perspective on Student Reasoning

Unlike the areas of developmental and cognitive psychology, student reasoning has not been a traditional focus of the research on science education with linguistically and culturally diverse student groups. Instead, much of this research has focused on issues such as ensuring equitable learning opportunities (Firestone, Camilli, Yurecko, Monfils, & Mayrowetz, 2000; Tate, 2001), closing the achievement gap (Norman, Ault, Bentz, & Meskimen, 2001; Rodriguez, 2001), or finding ways to make science education more appealing to students who may not readily embrace an identity as a science learner (Brickhouse & Potter, 2001; Seiler, 2001). This research has generally been guided by sociocultural theories borrowed from anthropology and sociology and has rarely focused on questions that are connected to discipline-specific content learning. Still, there are two strands of the research on student diversity in science education that seem most relevant to better understanding the reasoning of ELL students—what we will refer to as research on *continuity* and research on *context*. Some illuminating parallels can be drawn between these issues of continuity and context and the psychological perspectives of knowledge-lean experimentation strategy and knowledge-rich conceptual change.

The question of continuity in classroom research focuses on the connections (or the lack thereof) that learners perceive between the academic tasks they are asked to perform in school and their own developing epistemologies as learners. In other words, how well do normative science practices overlap with students' ways of making sense of the world, and how willing are students to modify their epistemologies to more closely align with normative science practices? The cultural congruence model of learning, for example, argues that students engage more fully in science learning when practices are not seen to be in conflict with their broader worldview (Aikenhead, 2001; Tharp, 1997) and when the communication and interaction patterns in the learning environment do not conflict with culturally accepted norms (Lee & Fradd, 1998). There have been ongoing debates about the degree to which culturally and linguistically diverse students' ways of knowing and talking are continuous or discontinuous with the ways of talking and knowing characteristic of normative science practices (Lee, 2002; Warren, Ballenger, Ogonowski, Rosebery, & Hudicourt-Barnes, 2001). This research on continuity shows some parallels to the research on knowledge-lean experimentation strategy, in that it shares a largely domain-general focus on beliefs and communication patterns that cut across science content areas.

While the research on continuity deals with worldviews and interaction patterns, the research on context addresses the links between classroom learning and knowledge and skills developed beyond the classroom. Much of this work falls under a "funds of knowledge" perspective, highlighting cultural, linguistic, and subject matter knowledge and skills developed and shared by individuals within their communities (González & Moll, 2002; González, Moll, & Amanti, 2005). Understanding students' funds of knowledge requires consideration of a range of learning environments in which students might gain content knowledge or language skills related to a topic, including knowledge constructed in the home (Barton, Drake, Perez, St. Louis, & George, 2004), in the community (Rahm, 2002), and in peer groups (Buxton, 2005), as well as knowledge constructed in the classroom. Research on context shares commonalities with the knowledge-rich research on conceptual change in developmental and cognitive psychology in that the focus tends to be on domain-specific content knowledge as it is developed in unique settings and then potentially mobilized in other settings that share the same domain-specific content.

Research on science learning with linguistically and culturally diverse student groups also provides important lessons when considering the reasoning of ELL students. First, the research on continuity points to the added complexity that must be accounted for when considering the goal of ELL students adopting scientific discourse patterns. In addition to the linguistic challenges of developing the academic language of science, there are epistemological issues involving the degree to which students feel comfortable presenting and representing a normative science worldview (Deyhle & Lecompte, 1994; Hammond, 2001). Teachers need to understand this issue and be willing and able to talk about it explicitly with students who are struggling with questions of belief as well as issues of content and language. Second, the research on context highlights the critical and multifaceted role that prior knowledge plays in developing scientific reasoning. ELL students are often viewed from a deficit perspective due to both their limited English proficiency and the perception (sometimes accurate and sometimes not) that immigrant ELL students bring limited formal education experiences from their countries of origin. Effective teachers of ELL students find ways to move beyond these perceived deficits to make connections between students' lived experiences and grade-appropriate content learning and cognitive complexity (Lee & Fradd, 1998; Warren, Ballenger, Ogonowski, Rosebery & Hudicourt-Barnes, 2001).

In summary, our reading of the research both in developmental and cognitive psychology and in studies of diverse learners points to parallels between these seemingly distant fields and a convergence around the value of teachers attending to their students' emerging scientific reasoning across school and home contexts. While our conceptual and analytical frameworks for fostering reasoning among culturally and linguistically diverse students continues to evolve, we believe that these frameworks will prove fruitful in promoting both science content learning and academic language development for ELL students.

APPLYING OUR REASONING FRAMEWORK IN PRACTICE

Methods

Grounded in our conceptual framework for understanding and supporting the scientific reasoning of culturally and linguistically diverse learners, we developed a series of three elicitation protocols to examine student reasoning about science tasks in both school and home contexts. One protocol was developed for each of the three grade levels (third through fifth) in the larger intervention project. Measurement was the topic selected for third grade, transfer of energy for fourth grade, and changing seasons for fifth grade. Here we will be discussing only the third- and fourth-grade reasoning data.

Third-grade students participated in the measurement elicitation task one-on-one with a member of the research team. The interviews were conducted soon after the students completed the measurement curriculum unit. All interviews were conducted in English, rather than in the students' home language, as English was the language of science instruction in all classes. The task was composed of four parts, with each part addressing four of the topical areas covered in the measurement curriculum unit: length, weight, volume, and temperature (see Appendix A for the elicitation protocol).

The first part of the interview asked students to discuss experiences they had with measurement in the home context. The second part asked students to perform measurement tasks using a ruler, kitchen scale, graduated cylinder, measuring cup, and thermometer. The third part asked students to make estimates about each of the four topical areas of measurement. The final part asked students to discuss experiences they had with measurement in the context of playing with their peers in out-of-school settings. Together, parts one and four were taken to represent the "home" context and parts two and three to represent the "school" context. Each part began with an opening question followed by one or more probes. The task took approximately 30 minutes and was videotaped. The videos were then transcribed for analysis.

The fourth-grade force and motion elicitation followed a parallel procedure, except that there were only three parts: part one asked students to discuss experiences they had with transfer of energy in the home context, part two had students design an experiment on transfer of energy, and part three asked students to discuss experiences they had with transfer of energy in the context of playing with their peers (see Appendix B for the elicitation protocol).

We developed an analytical and coding scheme to categorize student reasoning complexity. We constructed semantic maps of verbal and procedural responses and actions for each of the student interviews. These semantic maps provided an overall verbal and diagrammatic picture of each interview.

Each of the conceptual propositions was then scored using a reasoning complexity rubric along five criteria: (1) generativity (amount and types of topics and assertions brought forth by students); (2) elaboration (supporting details added to ideas within a topic); (3) justification (how well ideas are warranted with evidence and inference); (4) explanation (how underlying causal mechanisms are used to explain assertions); and (5) logical coherence (overall quality of connections between one assertion and other assertions) (see Table 10.1). The first two criteria assess the variety and richness of ideas raised by students, the third and fourth criteria assess the structure of student

TABLE 10.1: Student Reasoning Complexity Rubric

CRITERIA	0	1	2	3	4
Generativity	No observations	One or two observations or simple statements	Three or more observations or simple statements	One or two assertions or conjectures	Three or more assertions or conjectures
Elaboration	No elaboration	One or two elaborations of one idea	One or two elaborations of more than one idea	Three or more elaborations of one idea	Three or more elaborations of more than one idea
Justification	No justifications	One or two justifications of one assertion	One or two justifications of more than one assertion	Three or more justifications of one assertion	Three or more justifications of more than one assertion
Explanation	No explanations	Single mechanism explaining one assertion	Single mechanisms explaining more than one assertion	Multiple or chained mechanisms explaining one assertion	Multiple or chained mechanisms explaining more than one assertion
Logical Coherence	No logical connections or nonsensical connections	Vague connections making superficial sense	Clear and reasonable connections but lack support	Clear and reasonable connections with some specific supports	Clear and reasonable connections with many specific supports

TABLE 10.2: Reasoning Complexity for Entire Sample

N	GENERATIVITY M (SD)	ELABORATION M (SD)	JUSTIFICATION M (SD)	EXPLANATION M (SD)	LOGICAL COHERENCE M (SD)	DF	F	p
73	3.7 (0.5)	2.5 (0.6)	1.2 (0.6)	0.6 (0.5)	1.6 (0.4)	4	581.70	.001*

*Significant at $p = .001$.

reasoning, and the final criterion assesses the overall quality of student reasoning. Each of the five complexity criteria was scored on a 5-point scale (0–4), meaning that a score from a zero to four was assigned to each conceptual proposition map for each of the five criteria, and a composite score from a zero to 20 was assigned for overall reasoning complexity.

Finally, tables were constructed to look at cross-case analysis of the five criteria of reasoning complexity (generativity, elaboration, justification, explanation, and logical coherence). For third grade, cross-case analysis was also done for patterns in each of the four measurement topical areas (length, weight, volume, and temperature). In each case, the analysis was conducted with regard to three student demographic variables of ESOL level, home language, and gender. Key findings from analysis of the third-grade data will be presented first, followed by key findings from the fourth-grade data.

Findings

Third Grade

Analysis of overall reasoning complexity for the entire third-grade sample is presented in Table 10.2. The analysis indicates significant differences among the five criteria of reasoning complexity for the entire sample; $F(4) = 581.70$, $p < .001$. Analysis yielded a medium effect magnitude; $x^2 = 0.89$. Pairwise comparison t-tests indicated that each combination of pairs was statistically significant at $p < .001$. In other words, each reasoning complexity category (generativity, elaboration, justification, explanation, and logical coherence) is significantly different from the others. The result suggests that students were able to generate topics and examples involving measurement, but were less able to elaborate on those examples with supporting details, still less able to give rational justifications for their ideas, and the least able to give explanations based on underlying mechanisms.

Reasoning data were disaggregated by ESOL level, home language (Spanish or Haitian Creole), and gender. For ESOL level, we followed the school district framework of emergent English speakers (ESOL 1/2), developing English fluency (ESOL 3/4), conversationally fluent but still requiring academic language support (ESOL 5), and non-ESOL. No statistical significance was found for any of the criteria of reasoning complexity by ESOL level. Likewise, there were no significant differences by home language or gender. In other words, the same general pattern of reasoning was seen for third graders no matter how the demographic data were disaggregated.

When third grade reasoning was analyzed across the four topical areas of measurement (length, weight, volume, and temperature), there were no significant differences for the entire sample. However, there were differences when disaggregated by ESOL level. ESOL level 5 students scored significantly higher than ESOL level 1/2 and non-ESOL students for the topics of length and volume ($p = .05$). Third-grade topical area reasoning showed no significant differences by home language or gender.

Fourth Grade

For the fourth grade students, analysis of overall reasoning complexity for the entire sample is shown in Table 10.3. Results are quite similar to those of the third-grade sample, indicating significant differences among the five criteria of reasoning complexity for the entire sample; $F(4) = 245.45$, $p < .001$ and a medium effect magnitude; $x^2 = 0.75$. Pairwise comparison t-tests indicated that each combination of pairs was statistically significant at $p < .001$, with the exception of elaboration and justification. In other words, each of the reasoning complexity categories (generativity, elaboration, justification, explanation, and logical coherence) was significantly different from the others, except for elaboration and justification. The result suggests that students were able to generate topics and examples involving energy transformations, but were less able to elaborate on those examples with supporting details or to give rational justifications for their ideas, and the least able to give explanations based on underlying causal mechanisms.

TABLE 10.3: Reasoning Complexity for Entire Sample

N	GENERATIVITY M (SD)	ELABORATION M (SD)	JUSTIFICATION M (SD)	EXPLANATION M (SD)	LOGICAL COHERENCE M (SD)	DF	F	P
81	3.8 (0.4)	2.2 (1.0)	2.2 (0.7)	0.3 (0.6)	2.5 (0.9)	4	245.45	<.001*

* Significant at $p = .001$.

The fourth-grade reasoning data were disaggregated by ESOL level, home language (Spanish or Haitian Creole), and gender. Significant differences were observed in reasoning complexity by ESOL level, with ESOL level 5 students significantly outperforming the other ESOL levels, including non-ESOL students, for justification ($p = .05$). No significant differences were found by home language, but when disaggregated by gender, differences in the area of justification were found, with boys outperforming girls ($p = .05$).

DISCUSSION AND IMPLICATIONS FOR TEACHING SCIENCE TO ELL STUDENTS

Discussion

We have presented an overview of the findings from the third and fourth grade reasoning elicitation protocols as examples of how we have enacted our reasoning framework in practice. Our analytic framework for exploring the reasoning of ELL students was derived from our conceptual framework, highlighting the value of developmental and cognitive psychology perspectives as well as cultural and linguistic diversity perspectives to better understand how students think about science tasks.

Our findings proved to be intuitive in some ways and counterintuitive in other ways. For the third-grade students, we had hypothesized that the higher levels of reasoning complexity (justification and explanation) would be more challenging than the lower levels of reasoning complexity (assertions and elaborations). This pattern was clearly seen for the overall sample, with an inverse relationship between increasing level of reasoning complexity and student scores on the rubric.

When we disaggregated the third grade data by demographic subgroups, we hypothesized that the higher the students' ESOL levels were, the better they would score on the reasoning rubric, both because they had more English language ability with which to express themselves and because they would have been better able to comprehend the in-class instruction (primarily done in English) relevant to the elicitation questions. Contrary to our hypothesis, there were fewer and smaller differences than we had anticipated. This finding might be attributable to the fact that our intervention was targeted specifically to support science learning and English language development of ELL students, thus assisting the lower level ESOL students in performing only modestly below their more English proficient peers. Additionally, we were surprised by the finding that ESOL level 5 students outperformed everyone, including the non-ESOL students in topical reasoning. For the demographic subgroups of home language and gender in third grade, we hypothesized that there would be few, if any, significant differences; indeed, none were found.

For the fourth grade students, again we found a combination of intuitive and counterintuitive results. We hypothesized that as students both matured and spent more time in our intervention, the more challenging levels of reasoning complexity would increase. The scores for the lower reasoning levels of assertions and elaborations continued to be high. Fourth grade students did improve in their ability to provide justifications for their assertions. However, results continued to be low for the highest level of reasoning complexity, i.e., explanations of causal mechanisms. While students had studied the concepts that would serve as the underlying mechanisms for explaining transfer of energy, the curriculum provided few opportunities for students to explain these concepts in as loosely guided a fashion as was the case for the elicitation protocol. Students clearly did not interpret the questions in ways that prompted the activation of this prior knowledge.

When we disaggregated the fourth grade data by demographic subgroups, again we hypothesized that the higher the students' ESOL levels were, the better they would score on the reasoning rubric. Contrary to our hypothesis, we found a repeat of the pattern observed with the third-grade sample. Again, ESOL level 5 students outperformed everyone, including the non-ESOL students. While it might be expected that conversationally fluent ESOL level 5 students would outperform ESOL levels 1 to 4 students, we were surprised that the ESOL level 5 students also outperformed the non-ESOL students. The result suggests that ESOL level 5 students were both gaining the academic language of science in English and successfully employing this developing proficiency to execute and express their reasoning about science topics. Thus, an instructional intervention that is primarily in English but with robust supports for academic language development, such as our intervention, could be especially beneficial to students at the level of language proficiency typical of ESOL level 5 (Buxton, Lee, & Santau, 2008).

For the other demographic subgroups by home language and gender in fourth grade, we hypothesized that there would be few, if any, significant differences for the same reasons noted for third grade. For gender groups, we were somewhat surprised that there was a significant difference, with boys outperforming girls in the area of justification. We interpret that this could be, at least in part, a function of the topic of energy transfer. We purposefully selected topics that we believed were gender-neutral when we designed the elicitation task, but the fact remains that fourth-grade boys may be more likely to have experiences taking these things apart (e.g., flashlights, bicycles, soccer balls) and thinking about the processes involved in how they work.

These findings contribute to refining and verifying our conceptual framework for how the reasoning of linguistically and culturally diverse students develops. From the developmental and cognitive psychology perspective, the findings point to the roles of both a domain-generic experimentation strategy and a more domain-specific conceptual change. Part of what might have pushed the fourth-grade students to justify their assertions better than the third-grade students was the ongoing use of the Promoting Science among English Language Learners (P-SELL) curriculum that provided extensive opportunities to practice generating hypotheses, controlling variables and explaining evidence, all hallmarks of experimentation strategy research (Kuhn, 1990). At the same time, part of the challenge of why students so rarely generated examples of causal explanations might stem from the fact that the P-SELL curriculum, because it must cover all of the state science benchmarks, did not provide students with the depth of content knowledge on any given topic (e.g., transfer of energy) needed to promote rich conceptual change (diSessa & Sherin, 1998).

From the perspective of student diversity, these findings point to the important roles of both domain-generic cultural and linguistic continuity and more domain-specific knowledge in context. The fact that so many students seemed content to limit their reasoning to the lower complexity levels of assertions, with occasional elaborations, speaks to the question of whether students' ways of knowing and talking are continuous or discontinuous with the ways of talking and knowing characteristic of normative science practices (Lee, 2002). At the same time, part of what enabled ELL students, even those with emergent levels of English proficiency (i.e., ESOL levels 1 and 2), to generate high scores for the lower reasoning complexity levels might be traced to the focus in the P-SELL curriculum on identifying relevant prior knowledge both through formal science instruction and through funds of knowledge in their communities (González, Moll, & Amanti, 2005).

In short, the findings of this study support the assertion that for ELL students, as for all students, engaging in complex science tasks is essential for the development of scientific reasoning skills, which are supported by science content knowledge and academic English. Understanding and building upon students' prior knowledge, including connections to students' home language and culture, would seem to be an essential part of this approach.

Implications

The conceptual and analytical frameworks we have developed for the investigation of reasoning with culturally and linguistically diverse students are meant to serve as a foundation for our ongoing research and for other related studies. Our focus on ELL students' reasoning raises important questions about relationships between English language proficiency and reasoning complexity, as well as the role that teachers can play in supporting the development of their students' reasoning abilities.

We conceptualized the study of student reasoning as a way to get teachers to think about their students' prior knowledge both from developmental and from sociocultural perspectives. Thus, part of our larger research program examines teacher reasoning, as each teacher watches the videotape of the student from his/her classroom and reflects on what the student was thinking and why. The combination of student reasoning and teacher reasoning will offer multiple avenues for further studies. This approach of promoting teacher reasoning about student reasoning has been advocated in the past (Fennema & Franke, 1992), but has not been a focus of research in science education to date.

The results from this study have also provided new directions for our intervention efforts in teacher professional development and the development of improved curriculum materials for teaching science to ELL students. In our work with teacher professional development we have created multiple sessions on the value of attending to students' reasoning by focusing on the question of what students are thinking and why. We push teachers to think about where students' ideas come from and how these ideas develop from both formal and informal experiences. We see this as a back door approach to encouraging teachers to consider their students' funds of knowledge including their cultural and linguistic resources, areas that many teachers tend to deemphasize and undervalue.

It is clear from our findings that getting students to talk about science is important for the development of their academic language and scientific reasoning. However, not all kinds of science talk are equally valuable, if the goal is to push students towards expressing higher levels of reasoning complexity. Teachers who are consciously monitoring their students' talk for evidence of reasoning complexity can learn to promote science talk that focuses on justifications and causal explanations. While these higher levels of reasoning will remain challenging for elementary grade students, the latest synthesis of research on science learning (NRC, 2007) makes it clear that elementary students are capable of engaging in cognitively complex tasks. In our curriculum development efforts, we continue to build upon our findings from the student reasoning data. For example, in the fifth-grade curriculum, we have included longer instructional sequences focusing on collecting and evaluating evidence, opportunities to engage in model building activities, and supports for engaging in focused scientific conversations. We believe that such experiences will continue to assist ELL students in practicing more complex scientific reasoning.

We expect that our intervention to enhance student reasoning will lead to stronger outcomes in science learning and academic language development. Our ongoing research will look more directly at this relationship. We will also refine our conceptual and analytic frameworks for understanding the progression of reasoning with ELL students by leveraging new thinking from both cognitive and developmental psychology and research on student diversity as well as our own findings. We believe that this conceptual framework for supporting enhanced scientific reasoning for ELL students can aid us in our overarching goal of fostering both high academic achievement and educational equity with diverse student groups in science classrooms.

REFERENCES

Aikenhead, G. (2001). Cultural relevance: Whose culture? What culture? In J. Wallace & W. Louden (Eds.), *Dilemmas of science teaching* (pp. 92–95). New York: Routledge Falmer.

August, D., & Hakuta, K. (Eds.). (1997). *Improving schooling for language-minority children: A research agenda*. Washington, DC: National Academy Press.

Barton, A. C., Drake, C., Perez, J., St. Louis, K., & George, M. (2004). Ecologies of parental engagement in urban education. *Educational Researcher, 33*(4), 3–12.

Brickhouse, N., & Potter, J. (2001). Young women's scientific identity formation in an urban context. *Journal of Research in Science Teaching, 38*(8), 965–980.

Buxton, C. A. (2005). Creating a culture of academic success in an urban science and math magnet high school. *Science Education, 89*(3), 392–417.

Buxton, C., Lee, O., & Santau, A. (2008). Promoting science among English language learners: Professional development for today's culturally and linguistically diverse classrooms. *Journal of Science Teacher Education 19*(5), 495–511.

Carey, S. (1987). *Conceptual change in childhood* (1st MIT Press ed.). Cambridge, MA: MIT Press.

Cochran-Smith, M. (1995). Color blindness and basket making are not the answers: Confronting the dilemmas of race, culture, and language diversity in teacher education. *American EducationalResearch Journal, 32*(3), 493–522.

Deyhle, D., & LeCompte, M. D. (1994). Conflict over child development: Navajo culture and the middle schools. *Theory into Practice, 23*(3), 156–167.

diSessa, A. A., & Sherin, B. L. (1998). What changes in conceptual change? *International Journal of Science Education, 20*(10), 1155–1191.

Eisenhart, M., Finkel, E., & Marion, S. (1996). Creating the conditions for scientific literacy: A re-examination. *American Educational Research Journal, 33*(2), 261–295.

Fennema, E., & Franke, M. L. (1992). Teachers' knowledge and its impact. In Grouws, D. A (Ed.), Handbook of research on mathematics teaching and learning (pp. 147–164). New York: Macmillan Publishing Company.

Firestone, W. A., Camilli, G., Yurecko, M., Monfils, L., & Mayrowetz, D. (2000). State standards and opportunity to learn in New Jersey. *Education Policy Analysis Archives, 8*(35), 1–25.

González, N., & Moll, L. (2002). Cruzando el puente: Building bridges to funds of knowledge. *Educational Policy, 16*(4), 623–641.

González, N., Moll, L. C., & Amanti, C. (2005). *Funds of knowledge: Theorizing practices in households, communities, and classrooms.* Mahwah, NJ: L. Erlbaum Associates.

Hammond, L. (2001). Notes from California: An anthropological approach to urban science education for language minority families. *Journal of Research in Science Teaching 38*(8), 983–999.

Inhelder, B., & Piaget, J. (1964). *The early growth of logic in the child, classification and seriation.* New York: Harper & Row.

Keil, F. C., & Wilson, R. A. (2000). *Explanation and cognition.* Cambridge, MA: MIT Press.

Klahr, D. (2000). *Exploring science: The cognition and development of discovery processes.* Cambridge, MA: MIT Press.

Kuhn, D. (1991). *The skills of argument.* Cambridge: Cambridge University Press.

Kuhn, D. (1990). *Developmental perspectives on teaching and learning thinking skills.* Basel, NY: Karger.

Lehrer, R., & Schauble, L. (2005). Scientific thinking and science literacy. In W. Damon, R. Lerner, K. A., Renninger, & I. E. Sigel, (Eds.), *Child psychology in practice.* Hoboken, NJ: John Wiley & Sons.

Lee, O. (2002). Science inquiry for elementary students from diverse backgrounds. In W. G. Secada (Ed.), *Review of research in education, Vol. 26* (pp. 23–69). Washington, DC: American Educational Research Association.

Lee, O., & Fradd, S. (1998). Science for all, including students from non-English-language backgrounds. *Educational Researcher, 27*(4), 12–21.

Metz, K. E. (2004). Children's understanding of scientific inquiry: Their conceptualization of uncertainty in investigations of their own design. *Cognition and Instruction, 22*(2), 219–290.

Metz, K. E. (1991). Development of explanation: Incremental and fundamental change in children's physics knowledge. *Journal of Research in Science Teaching, 28*(9), 785–797.

National Assessment of Educational Progress (NAEP) 2005. Accessed March 26, 2010 at http://www.nationsreportcard.gov/science_2005

National Center for Education Statistics. (1999). *Teacher quality: A report on the preparation and qualifications of public school teachers.* Washington, DC: U.S. Department of Education, Office of Educational Research and Improvement.

National Research Council. (2007). *Taking science to school.* Washington, DC: National Academies Press.

Norman, O., Ault, C. R. Jr., Bentz, B., & Meskimen, L. (2001). The black-white "achievement gap" as a perennial challenge of urban science education: A sociocultural and historical overview with implications for research and practice. *Journal of Research in Science Teaching, 38*(10), 1101-1114.

Piaget, J. (1973). *The child and reality: Problems of genetic psychology*. New York: Grossman Publishers.

Rahm, J. (2002). Emergent learning opportunities in an inner-city youth gardening program. *Journal of Research in Science Teaching, 39*(2), 164-184.

Rodriguez, A. (2001). From gap gazing to promising cases: Moving toward equity in urban education reform. *Journal of Research in Science Teaching, 38*(10), 1115-1129.

Schauble, L. (1996). The development of reasoning in knowledge-rich contexts. *Developmental Psychology, 32*(1), 102-119.

Seiler, G. (2001). Reversing the "standard" direction: Science emerging from the lives of African American students. *Journal of Research in Science Teaching, 38*(9), 1000-1014.

Smith, C. L., Maclin, D., Houghton, C., & Hennessey, M. G. (2000). Sixth-grade students' epistemologies of science: The impact of school science experiences on epistemological development. *Cognition and Instruction, 18*(3), 349-422.

Tate, W. (2001). Science education as a civil right: Urban schools and opportunity-to-learn considerations. *Journal of Research in Science Teaching, 38*(9), 1015-1028.

Teachers of English to Speakers of Other Languages. (1997). *ESL standards for pre-K-12 students*. Alexandria, VA: Author.

Teachers of English to Speakers of Other Languages. (2006). *Pre-K-12 English language proficiency standards*. Alexandra, VA: Author.

Tharp, R. (1997). *From at-risk to excellence: Research, theory, and principles for practice*. Santa Cruz, CA: Center for Research on Education, Diversity & Excellence.

Toth, E., Klahr, D., & Chen, Z. (2000). Bridging research and practice: A cognitively based classroom intervention for teaching experimentation skills to elementary school children. *Cognition and Instruction, 18*(4), 423-459.

Warren, B., Ballenger, C., Ogonowski, M., Rosebery, A., & Hudicourt-Barnes, J. (2001). Re-thinking diversity in learning science: The logic of everyday language. *Journal of Research in Science Teaching, 38*(5), 529-552.

Wong-Fillmore, L., & Snow, C. (2002). *What teachers need to know about language*. Washington, DC: Center for Applied Linguistics.

Zimmerman, C. (2000). The development of reasoning skills. *Developmental Review, 20*(1), 99-149.

APPENDIX A

Grade 3 Student Reasoning Elicitation Protocol: Measurement

Opening: Tell the student, "We are going to do a science activity today about measuring things. I'm going to ask you some questions to see what you know about measurement. This isn't a test and I don't expect you to know all the answers. I just want to know how a smart kid like you thinks about these things. Ok? You can ask me any questions about what we are doing whenever you want to. Ok?"

Section 1: Home Context

The purpose of the first section of the interview is to get the student to reflect on ways that he/she has engaged in measurement tasks (or observed others engage in measurement tasks) in the home context.

Initial prompt: We are going to start by talking about times when you measure things at home.

1. Have you ever seen anyone in your family measure how long or how tall something is? What were they measuring? How did they do it? What did they do it with?
   Probe: How tall are you? How do you know?

2. Have you ever seen anyone in your family weigh something to see how heavy it is? What were they weighing? How did they do it? What did they do it with?
   Probe: Who does most of the cooking in your family? Have you ever seen them weigh anything when they are cooking? What did they weigh? How did they weigh it? Do you ever go to the food store (market)? Have you ever seen anyone weigh anything there? What did they weigh? How did they weigh it?

3. Have you ever seen anyone in your family find the volume or capacity of something to see how much space it takes up? What were they finding the volume of? How did they do it? What did they do it with?
   Probe: Have you ever needed to take medicine when you were sick? Who gave it to you? How did they know how much to give you? How did they do this?
   Have you ever helped do the laundry? How do you know how much detergent (soap) to put into the washing machine?

4. Have you ever seen anyone in your family take the temperature of anything? What were they taking the temperature of? How did they do it? What did they do it with?
   Probe: Have you ever had a fever? Did someone find out what your temperature was? How did they do this?
   Have you ever heard people ask about the temperature outside? How do they find out what it is?

Section 2: School Context: Tool Use

Now I'm going to ask you to measure some things for me using some tools. This may be like the kind of activity you are doing in science time here in school.

First, I'd like you to take this plastic cup and fill it about half way up with water from this container.

Is that about halfway full?

1. Now can you tell me the height of the water in the cup? Which tool do you want to use to find out how high the water is in the cup? Ok, go ahead and do it. Can you write your answer in the box for length? (point to place on student sheet)
   (*Note:* for all of the measurements, the tools have both metric and traditional systems. Let the student measure in whichever system he/she wishes. If the student asks you which system to use, tell him/her to use the one he/she prefers.)
2. Now I'd like you to weigh the water in the cup. Which tool do you want to use to find out how heavy the water is in the cup? Ok, go ahead and do it. Can you write your answer in the box for weight? (point to place on student sheet)
   (*Note:* if student struggles with the units on the scale, prompt with "Do you know how many grams are in a kilogram or how many ounces are in a pound? Can you find 1 kilogram or 1 pound on the scale? What do you think the 200, 400, 600, that go up to 1 kg or the 4, 8, 12 that go up to 1 lb. stand for? Now can you tell me how heavy the water in the cup is?")
3. Now can you tell me the volume or capacity of the water in the cup? (Prompt: remember that means how much space it takes up.) Which tool do you want to use to find out how much space the water in the cup takes up? Ok, go ahead and do it. Can you write your answer in the box for volume? (point to place on student sheet)
4. Last, I'd like you to tell me the temperature of the water in the cup. Which tool do you want to use to find out how hot or cold the water in the cup is? Ok, go ahead and do it. Can you write your answer in the box for temperature? (point to place on student sheet)

	LENGTH (HEIGHT)		VOLUME (CAPACITY)	
	centimeter (cm)	inch (in)	milliliter (mL)	ounce (oz)
Cup half-full				
	Weight		Temperature	
	gram (g)	ounce (oz)	Celsius (°C)	Fahrenheit (°F)

Section 3: School Context: Comparing and Estimating

Now take this second plastic cup just like the first one. This time, I'd like you to fill it completely with water, all the way to the top.

Now I want you to compare the water in the half-filled cup with the water in the full cup.

I have some questions for you to answer. These are sentences that have multiple-choice answers and I want you to tell me which is the best answer. Can you read the first sentence to me?

1. The height of water in the full cup is higher than, lower than, or the same as the height of water in the half-filled cup.

    higher　　　lower　　　same

    Remember when you measured the height of the water in the half cup you said it was [whatever student wrote down]. Now, without measuring, what do you think the height of the water in the full cup is?

    _____

Now can you read me number 2? It's another multiple-choice sentence.

2. The weight of the water in the full cup is heavier than, lighter than, or the same as the weight of water in the half-filled cup.

    heavier　　　lighter　　　same

    Remember when you weighed the water in the half cup you said it was [whatever student wrote down]. Now, without measuring, what do you think the weight of the water in the full cup is?

    _____

Now can you read me number 3.

3. The volume of water in the full cup is larger than, smaller than, or the same as the volume of water in the half-filled cup.

    larger　　　smaller　　　same

    Remember when you measured the volume of the water in the half cup you said it was [whatever student wrote down]. Now, without measuring, what do you think the volume of the water in the full cup is?

    _____

Now can you read me number 4?

4. The temperature of water in the full-cup is warmer than, colder than, or the same as the temperature of water in the half-filled cup.

    warmer　　　colder　　　same

Remember when you took the temperature of the water in the half cup you said it was [whatever student wrote down]. Now, without measuring, what do you think the temperature of the water in the full cup is?

_____

Section 4: Connection to Play

So, we talked about measuring and estimating today. We practiced using tools like the ruler and the thermometer. Now, the last thing I want to ask you about is if you ever use measurement when you are playing with your friends or with your brothers or sisters.

1. Can you think of a time playing with your friends that you might need to measure something? Tell me about it.
   Probe: Have you ever wanted to figure out how far something goes? How heavy something is?
2. In the example you were just talking about of [give a student example], do you think it would make a difference if you measured using a tool like the ruler or the scale, or would estimating be good enough? Why do you think so?
3. The tools we used today had two different systems, the metric system (centimeters, grams, milliliters, and Celsius degrees) and the traditional system (inches, ounces, cups, and Fahrenheit degrees). Which of these two systems is easier for you to use? Why do you think so?

APPENDIX B

Grade 4 Student Reasoning Elicitation Protocol: Forms of Energy

Opening: Tell the student, "We are going to do a science activity today about forms of energy and rolling balls. This isn't a test and I don't expect you to know all the answers. I just want to know how a smart kid like you thinks about these things. Ok? You can ask me any questions about what we are doing whenever you want to. Ok?"

Section 1: Home Context

[The purpose of the first section of the interview is to get the student to reflect on ways that he/she has experienced various forms of energy in the home context.]

Initial prompt: We are going to start by talking about things that use energy at home.

1. Can you tell me all the things that use energy in your house?
   Probe: Can you think of other things in your house that use energy? Can you give me any more examples?

2. Sometimes energy is converted from one form to another. Think about a light bulb. What kind of energy does it need to work? What kind of energy is the result?
   Probe: Can you describe what is happening with the light bulb?
3. Think about a flashlight. What kind of energy does it need to work? What kind of energy is the result?
   Probe: Can you describe what is happening with the flashlight?
4. Think about a washing machine. What kind of energy does it need to work? What kind of energy is the result?
   Probe: Can you describe what is happening with the washing machine?

Section 2: School Context: Potential and Kinetic Energy

Tell the student, "Now I'm going to ask you to do a science experiment. This may be like the kind of activity you are doing with your teacher during science here in school. First, I'd like you to look at the inquiry framework in this handout."

Have you used this before in class?
Probe: Tell me about when you used it?
What do you remember about using it?
Today, we are going to use the inquiry framework to think about forms of energy.

INQUIRY FRAMEWORK

1. **Questioning**

   This is the question we want to answer today:
   Suppose you roll a ball down a ramp. How does changing the weight of a ball change the amount of energy it has?
   (Have the student read the hypotheses. If the student struggles or appears puzzled, the interviewer should re-read the hypotheses to the student.)

   Make a hypothesis
   The lighter the ball is, the more energy it has.
   The heavier the ball is, the more energy it has.
   Changing the weight of the ball does not affect the energy it has.
   There is not enough information to tell.

2. **Planning**

   Now we will make a plan:
   You have a ramp to roll the balls down and a cup to roll the balls in to. You also have 3 balls of different weights.

Remember what you want to find out: *How does changing the weight of the ball change the amount of energy it has?*
Tell me your plan for what you will do.
Probe:
Is there something you can measure to help you answer this question?
How many times should you roll each ball?
If you roll a ball once or 3 times, which will give you a better answer? Why?
What if you make a mistake or something goes wrong?
Where will you put the ball on the ramp to start it?
Where will you place the cup before you roll the ball?
(*Ask this question after first trial) Where will you start measuring from? Where will you measure to?

3. **Implementing**

   Gather the materials
   1 ramp
   1 meter stick
   3 balls (differing weights but all other variables remain the same)
   1 box (to set ramp height)
   1 cup (cut in half to receive ball)

   Remember what you said you would do. Remember to write down all your measurements in the data table.

	BALL #1 11 GRAMS	BALL #2 22 GRAMS	BALL #3 53 GRAMS
Trial 1			
Trial 2			
Trial 3			

4. **Concluding**

   Draw a conclusion
   What did you find out? Check the most correct statement based on your data.
   - The lighter the ball was, the farther it pushed the cup.
   - The heavier the ball was, the farther it pushed the cup.
   - Changing the weight of the ball did not affect how far the ball pushed the cup.
   - There is not enough information to tell.

(*Only if student answers "There is not enough information to tell," ask, "What additional information would you need to answer the question?")

What can you conclude?
- The lighter the ball was, the more energy it had.
- The heavier the ball was, the more energy it had.
- Changing the weight of the ball did not affect the energy it had.
- There was not enough information to tell.

Compare what you thought would happen with what actually happened. Did the results match your hypothesis?
- Yes
- No

5. **Reporting**

   Share your results
   What happened when you changed the weight of the ball?
   Probe:
   How did changing the weight of the ball change the distance the cup moved?
   Why do you think this happened?
   Probe:
   Did the changing distance have anything to do with forms of energy? Why?
   Did energy get converted from one form to another?
   What kind of energy did the balls have at the top of the ramp?
   What kind of energy did the balls have at the bottom of the ramp?

Section 3: Connection to Play

Today we talked about different forms of energy at home. Then we did an experiment about energy and rolling balls. The last thing I want to ask you about is forms of energy when you are playing with your friends.

1. First, tell me all the things you like to play that use energy.
   Probe: Can you think of any more things you play that use energy? Can you think of any more examples?
2. When you slide down a sliding board on a playground, what kind of energy do you need to slide? What kind of energy is the result?

Probe: Are there similarities between you sliding down the sliding board and the balls rolling down the ramp? Tell me about it.

Suppose two children are different weights; one is heavy and one is light. When both slide down the sliding board, do they have the same amount of energy? Tell me about it.

Do they have the same forms of energy? Tell me about it.

3. When you kick a soccer ball, what kind of energy do you need to do it? What kind of energy is the result?

   Probe: Can you describe what is happening when you kick the ball?

4. When you ride a bike, what kind of energy do you need to do it? What kind of energy is the result?

   Probe: Can you describe what is happening when you ride the bike? Do you wear any protection when you bike? Do you think you should? Why?

READING 11

A NEGLECTED DIMENSION OF SOCIAL JUSTICE

A MODEL FOR SCIENCE EDUCATION IN RURAL SCHOOLS

By Mary John O'Hair and Ulrich C. Reitzug

One-third of all U.S. school children attend school in rural settings. Rural America is often much poorer than urban America, with most of the poorest counties in the United States located in rural areas. Equity is a concern not only in terms of race, class, gender, disability, and sexual orientation, but also in terms of being geographically located in a rural area. Rural teachers are often not certified in their teaching areas, with, for example, one in four rural science teachers lacking in academic preparation or certification. This chapter describes the K20 Oklahoma Science Initiative for Rural Schools that targets low-income, rural schools serving diverse populations in Oklahoma. The K20 Initiative helps reduce the professional, cultural, and social isolation and lack of professional development in rural schools. The objectives of the initiative are to improve teacher quality and student success through three research-based strategies which are described in the chapter.

Rural America, representing one-third of all U.S. schoolchildren, is much poorer than urban America, with 59 of the 66 poorest counties located in rural areas (Gates, 2004). Rural school districts, serving nearly 10 million children (Johnson & Strange, 2007) are at a disadvantage when competing for resources for professional development and attracting qualified teachers, with one in four rural science teachers lacking in academic preparation or certification (National Science Board [NSB], 2006). More than 400,000 educators teach in rural schools, representing 31% of all public school teachers (National Center for Education Statistics 2002). Compared to their nonrural counterparts, rural teachers average 13.4% less in salary, live in substandard housing, experience professional, cultural, and social isolation, and receive little if any professional development (Beeson & Strange, 2003; Darling-Hammond, 2000; Education Trust, 2003; Jimerson, 2003). Rural principals and superintendents feel ill-prepared for challenges that face them (Lamkin, 2006). Thus, although social justice is often discussed in terms of race, class, gender, disability, and sexual orientation, it may also be an issue of location—in this case, being located in a rural area.

The purpose of this chapter is to call attention to a neglected dimension of social justice—social justice for rural schools and, particularly, for the education of the students who attend these schools and the professional development of the educators who serve in them. We do this by describing the kindergarten through graduate education (K20) Oklahoma Science Initiative for Rural Schools, a program within the K20 Center for Educational and Community Renewal at the University of Oklahoma. The Oklahoma Science Initiative for Rural Schools [K20 SCIENCE] targets low-income, rural schools serving diverse populations in Oklahoma, including the 22,000 Native Americans who attend rural Oklahoma schools. K20 SCIENCE is one initiative that helps reduce the professional, cultural, and social isolation and lack of professional development in rural schools. The initiative is focused on science education due to the dire need in that area. The NSB (2006) reports that "the critical lack of technically trained people in the United States can be traced directly to poor K–12 mathematics and science instruction" (p. 2).

Research on professional development finds that teacher learning is greater when professional development utilizes an embedded professional development approach, linked directly to student achievement (e.g., lesson study, authentic research experience for teachers, professional learning communities), rather than the traditional workshop or conference format (Garet, Porter, Desimore, & Birman, 2001). Similarly, Fullan (2001, 2003) notes that to significantly improve student learning, teachers must be learning. Based on this knowledge, the K20 Center at the University of Oklahoma and the Oklahoma Science Project (OSP) developed an embedded professional development model. For 12 years, OSP has provided authentic research experiences for rural science teachers with 48 teachers (4 per year) completing the program. The K20 Center for Educational Renewal and Community Development has, for 10 years, promoted systemic "whole school" reform through a school-university network designed to transform conventional schools into professional learning communities (PLCs) using peer coaching, regional networking, and the IDEALS systemic change framework[1] (O'Hair, McLaughlin, & Reitzug, 2000). In recent years, over 500 rural schools have participated in this effort.

This chapter describes a professional development model that moves beyond a conceptual framework to one that is evidence-based and which combines strengths of OSP and K20 to promote exemplary science instructional practices in seventh–twelfth grade rural classrooms.

THE K20 SCIENCE INITIATIVE FOR RURAL SCHOOLS

The objectives of K20 SCIENCE are to improve teacher quality and student success through two main components: interactive instruction and embedded professional development. Merging these well-established approaches, the K20 SCIENCE Initiative for Rural Schools deepens teachers' content knowledge while impacting large numbers of rural teachers and classrooms across the state.

Interactive Instruction

The most prominent theories on how students develop understanding are based on the ideas that learning is active (Bransford, Vye, Bateman, & Brophy, 2004), involves the acquisition of organized knowledge structures and social interaction (Piaget, 1972; Vygotsky, 1978; Greeno, 1997), and relates new information to existing cognitive structures in order for learning to be meaningful (Blumenfeld, Soloway, Marx, & Krajcik, 1991;

Good & Brophy, 2000; Hannafin & Land, 1997; Jonassen, 1999). Additional research has documented substantial achievement benefits for *all* students, regardless of school level, size, context, ethnicity, or socioeconomic status (SES), when students are exposed to the kinds of teaching characterized as interactive instruction producing authentic intellectual work (Smith, Lee, & Newmann, 2001). This type of teaching results in students producing intellectual artifacts that are worthwhile, significant, and meaningful, such as those undertaken by successful adults (i.e., scientists and other professionals) who apply basic skills and knowledge to complex problems (Newmann, 1996; Newmann & Wehlage, 1995). Educational researchers have developed instructional strategies based on experiential learning, meaningfulness, and reflection, in order to facilitate the development of knowledge that can be applied more flexibly to different contexts and problems (e.g., Blumenfeld et al., 1991; Bransford, Brown, & Cocking, 2000).

Authentic assignments using interactive instruction require students to (a) construct knowledge involving organizing, interpreting, evaluating, or synthesizing prior knowledge to solve new problems; (b) engage in disciplined inquiry (i.e., use of a prior knowledge base; striving for in-depth understanding rather than superficial awareness; and expressing one's ideas and findings through elaborated communication); and (c) provide value beyond school for the learning. "These three criteria—construction of knowledge, through the use of disciplined inquiry, to produce discourse, products, or performances that have value beyond school—form the foundation for standards to assess the intellectual quality of teaching and learning" (Newmann, Bryk, & Nagaoka, 2001, p. 14). Recent research supports that interactive instruction producing more authentic intellectual work improves student scores on conventional tests (Newmann et al., 2001), student motivation to learn (Greene, Miller, Crowson, Duke, & Akey, 2004; Roeser, Midgley, & Urdan, 1996), and is linked to student success in high school science and mathematics (Lee, Croninger, & Smith, 1997). The effectiveness of interactive methods is supported by substantiated theory on how students learn (Bransford et al., 2000; Good & Brophy, 2000; Hannafin & Land, 1997; Jonassen, 1999).

Concerns about the reluctance of teachers to implement interactive instruction are not new (e.g., Blumenfeld et al., 1991). For students to receive interactive instruction that engages them in authentic intellectual work, teachers must learn new teaching methods and acquire more subject matter knowledge as well. After studying 2017 assignments from 277 teachers, Newmann and colleagues (2001) concluded that asking teachers to effectively implement authentic intellectual work necessitates providing resources for integration and assessment and professional networking opportunities.

The K20 SCIENCE model, described next, combines three strategies linked to accelerating and supporting change from didactic to interactive pedagogy: authentic research experiences for teachers, lesson study, and professional learning communities.

Embedded Professional Development

The embedded professional development component of K20 SCIENCE is grounded in three research-based strategies:

1. Deepening the content knowledge and comfort with inquiry-based teaching of rural secondary science teachers through *authentic research experiences*. Authentic research experiences for teachers deepen understanding of scientific inquiry while enriching substance and process (Kincheloe, 1991; Newmann, 2000).
2. Transferring and sustaining teachers' authentic research experiences into classroom practice through *lesson study* (credited with Japan's evolution in mathematics and science teaching; National Research Council [NRC], 2002).
3. Creating PLC that provide meaningful learning experiences for teachers and students. Professional learning communities, including peer coaching and regional networking, reduce the remoteness and isolation which affects rural teachers' learning (Malhoit & Gottoni, 2003) and support authentic intellectual work which has been associated with higher levels of student achievement (Lee & Smith, 1996).

The goals of K20 SCIENCE address specific needs through research-based strategies emphasizing interactive instruction.

Goal 1: Deepening Content Knowledge and Comfort With Inquiry-Based Teaching of Rural Secondary Science Teachers Through Authentic Research Experiences

Need

Research connects increased student achievement in science with teaching for understanding of both science disciplinary content *and* the centrality of inquiry in science (Pasley, Weiss, Shimkus, & Smith, 2004). Developing high quality mathematics and science teaching requires deepening teachers' content knowledge through sustained professional development (NSB, 2006). The National Science Education Standards identify more attention to inquiry as the hallmark of good science instruction (NRC, 2002). Teachers who have not experienced inquiry-based, interactive instruction are ill-equipped to employ this instructional strategy in their classrooms (Newmann, King, & Youngs, 2000).

Research-Based Strategy: Authentic Research Experiences for Teachers

Authentic research experiences for teachers provide sustained opportunities for teachers to experience an instructional strategy while they study, experiment with, and receive helpful advice on scientific content; collaborate with professional peers both within and outside of their schools; have access to external experts (i.e., research scientists); and have influence on both the substance and the process of their professional development (Newmann et al., 2000). Experiences of the OSP program over the past 11 years found that teachers readily gained the essentials of scientific inquiry, including confidence in their ability to carry through a rationally conceived research project from beginning to end, when provided authentic research experiences and guidance (Silverman, 2003). In addition to confidence in research ability, teachers' mastery of the scientific content increased significantly through the research experience (Slater & Cate, 2006). As an added bonus, mentor research scientists improved their teaching and communication skills through authentic pedagogy that encouraged critical reflection and knowledge construction through social interaction (Tanner, Chatman, & Allen, 2003).

Implementation
The K20 SCIENCE Summer Research Institute (SRI) engages teachers in scientific discovery with scientists from the University of Oklahoma and assists them in teaching for conceptual understanding. The SRI involves an in-depth, 5-week summer research experience for rural teachers, including:

1. An introduction to scientific research, research methods, and experimental materials.
2. The designing and conducting of original authentic research with guidance from scientists.
3. The translation of research experiences into classroom practices that focus on conceptual understanding.

Teachers utilize wireless-enabled laptop computers and scientific probes to gather, record, and analyze data. During and following the SRI, teachers work in lesson study teams (see Goal 2 discussion below). Upon completing the SRI, teachers submit a summary presentation of their research, reflections on the SRI, and formal lesson plans generated by lesson study teams. The K20 SCIENCE network is available for teachers to seek assistance and advice from other teachers, K20 SCIENCE staff, and research scientists. In addition, teachers completing the SRI have the opportunity to return as mentors for subsequent SRIs in order to support new teacher participants and add to their own research experience.

Goal 2: Transferring and Sustaining Teachers' Authentic Research Experiences Into Classroom Practice Through Lesson Study
Need
For 8 years, researchers at the Wisconsin Center for Education Research (WCER) National Center for Improving Student Learning and Achievement in Mathematics and Science (NCISLA) have worked with teachers and schools to create and study classrooms in which compelling new visions of mathematics and science are the norm. NCISLA found that fundamental reforms in learning and teaching are most likely to be achieved through professional development grounded in teacher inquiry and student conceptual under-

standing. Huffman and Hipp (2003) similarly found that teachers transfer their learning to the classroom and ultimately to student learning when they network with each other using processes, such as lesson study, designed to create mutual respect and trustworthiness among staff members.

Research-Based Strategy: Lesson Study

Lesson study originated in Japan and has been credited with Japan's evolution of effective mathematics and science teaching (Lewis, 2002a, 2002b; Lewis & Tsuchida, 1997; National Research Council, 2002). Lesson study is an iterative process focusing on what teachers want students to learn rather than on what teachers plan to teach (Lewis, 2002a; Stigler & Hiebert, 1999; Yoshida, 1999). A group of teachers develop a lesson together; one group member teaches the lesson while the others observe student learning; the group reconvenes to debrief, analyze, and if needed revise the lesson to incorporate the observations; and the teaching process begins again with a new teacher. This process of inquiring about their own teaching permits teachers to examine and adapt their practice, resulting in authentic achievement for students (Stewart & Brendufer, 2005). Through creating a culture of inquiry and demanding rigorous work, lesson study provides the opportunity for lifelong learning by teachers and a model for the students (Chokshi & Fernandez, 2004).

Implementation

Using the process of lesson study, the SRI staff guides and assists participants in integrating scientific research processes into classroom practices whose objective is teaching for conceptual understanding. Teachers collaborate (during and after their SRI experience) to craft and refine lessons that utilize the principles of inquiry they have practiced during their research experience. With support from the K20 SCIENCE staff, teachers work with their teams throughout the school year to reflect on, revise, and refine the lessons and instructional strategies from the SRI experience.

The purpose of lesson study is not simply the improvement of a single lesson. Rather lesson study provides teachers with an opportunity to examine their teaching in a way that results in the transfer of new knowledge acquired during the SRI research experience directly to their classrooms, ultimately resulting in the improvement of student achievement (Lewis, Perry, & Hurd, 2004). Teachers completing the SRI experience also engage in lesson study to share their learning with colleagues in their schools. Lesson study teams of four to six teachers, either within a school or within a region, meet for a full day monthly, using release time or stipends, to cooperatively build and script a science lesson and carry out the lesson study process. K20 SCIENCE staff members or outside experts are available during these meetings in a consulting capacity. During the lesson study process, the teachers collaborate to consider core and cross curricular strategies, analyze student learning, and develop small communities of practice. A teacher subsequently teaches the lesson to his/her own class, while the other team members (and K20 SCIENCE staff) carefully observe how the students are learning the concepts and skills. Following the lesson, observers report to the teacher and the team revises the lesson consistent with the feedback. Subsequently the revised lesson is re-taught by a different teacher. The debriefing and revision process continues until the teachers are satisfied with the lesson and feel that it exemplifies inquiry learning standards and maximizes student conceptual understanding.

Goal 3: Creating PLCs That Provide Meaningful Experiences for Learners

Need

An early OSP evaluation indicated the lack of sustaining variables such as teacher networking throughout the school year to deepen learning and accelerate the change process in science classroom practices (McCarty, 2003). In that program, teachers generated curriculum documents at the conclusion of their research experience that were posted on the OSP Web site. Subsequent interviews with OSP teachers revealed that these documents were rarely used after the initial SRI experience (Slater & Cate, 2006). Teachers indicated that although the curriculum documents were important, isolation and lack of peer interaction and support reduced their use of the documents.

PLCs

Researchers (Atkinson, 2005; Williams, 2006) have found that professional learning communities and technology integration (TI) provide supportive conditions that foster peer interactions and changes in classroom practices leading to interactive teaching that enhances authentic intellectual work of students. Early constructivist research (Dewey, 1938; James, 1958) supports the work of recent theorists (Dufour, Eaker, & Dufour, 2002; Hord, 1997; Senge, 2000), who report that PLCs, an approach to engaging school staffs in meaningful learning, leads to increased student achievement (Huffman & Hipp, 2003; Lee & Smith, 1996). Research reveals that a strong sense of community not only increases persistence, but also enhances information flow, learning support, group commitment, collaboration, and learning satisfaction (Rovai, 2002; Wellman, 1999). How successfully a science innovation travels across diverse conditions and geographical areas depends on the extent to which a teachers' academic support network is established (Carpenter, 2004). Technology helps expand teachers' access to a larger community of learners, and as noted by the National Staff Development Council (2001), allows teachers to exchange ideas with each other and leading experts in their content areas, visit classrooms of exemplary teachers, receive coaching from their mentors via internet conferencing, and access online virtual libraries (Loucks-Housley, Love, Stiles, & Mundry, 2003). Schools functioning as PLCs promote collective responsibility for student learning, develop norms of collegiality among teachers, and have been associated with higher levels of student achievement (Lee & Smith, 1996; Little, 1993; Louis, Marks, & Kruse, 1996). Not only is students' achievement significantly higher in schools that function as professional learning communities, but those gains also are also distributed more equitably. That is, the achievement gap between students of lower SES and students of higher SES is narrower (Lee & Smith, 1994). Developing PLCs and networks among rural educators decreases the remoteness and isolation that often plagues rural teachers (Malhoit & Gottoni, 2003) and improves rural teacher quality and student success while ensuring sustainability beyond direct K20 Center involvement (Hamlin, 2007; O'Hair, Williams, Wilson, & Applegate, in press). As rural schools incorporate characteristics of professional learning communities, a commensurate increase of TI into their practices, including student use for learning, has been documented (Atkinson, Williams, O'Hair, & O'Hair, 2008). Research findings suggest that this kind of organizational learning with continual renewal increases professional learning community development and school capacity to support and sustain change (Dufour et al., 2002; Hord, 1997; Huffman & Hipp, 2003; O'Hair, 2008; Sergiovanni, 1994).

Over the past 4 years, the state of Oklahoma's barometer for academic success (Academic Performance Index—API) has recorded significant improvement in K20 partner schools (74% higher than the state's average increase; Williams, Atkinson, O'Hair, & Applegate, 2007). Although other factors may impact API gains, the only attribute shared by participating schools is their involvement in the K20 Center's systemic reform model. These results may well extend into higher education and industry as students complete a multitude of degrees, particularly STEM-related, and enter the work force. K20 SCIENCE efforts support a strong foundation designed to support and nurture rural schools and to respond to "America's urgent challenge to remain a world leader in science and technology" (NSB, 2006, p. 1).

Implementation

A professional learning community can be as small as an individual classroom or as broad as a network of schools across a state. Within a classroom, the PLC is focused on effectively implementing K20 SCIENCE instructional strategies to increase student achievement in science and includes interactive instruction and authentic intellectual work standards (Newmann, 1996; Smith et al., 2001). Classroom practice considers students' prior knowledge and encourages construction of new knowledge based on experiments, demonstrations, extensive written and oral communication, problem-solving, and real world connections. Authentic, high-quality science lessons provide opportunities for students to interact purposefully with science content and focus on the overall learning goals of the concept.

At the school level, the PLC strategy expands learning of teachers to include the entire school community, creating a collaborative, sustaining culture to improve the school's capacity to help all students learn at high levels (Dufour et al., 2002). K20 SCIENCE teachers equipped with inter-net based video-conferencing capabilities are encouraged to network and collaborate on classroom projects with other rural teachers and scientists. Video-conferencing collaborations have received exceptionally positive feedback from students and teachers.

Additional opportunities for networking include the K20 network of 500 rural school leaders and their school boards. K20 SCIENCE strategies are introduced, studied, and discussed in initial 2-day leadership seminars for administrators wishing to join the K20 schools network; professional meetings with partner organizations such as the state's school boards association, administrator and teacher associations; and K20 regional cluster meetings designed for ongoing professional development.

CONCLUSIONS AND IMPLICATIONS

Low-income, rural students and their teachers are the forgotten underrepresented group. Students living in rural areas of the U.S. achieve at lower levels and drop out of high school at higher rates than their non-rural counterparts (Roscigno & Crowley, 2001). Additionally, in Oklahoma, the state in which the K20 SCIENCE model is based, the poverty rate substantially exceeds the national poverty level (Bishaw & Stern, 2006). Nearly 25% of Oklahoma students drop out of high school between ninth and twelfth grade and too many Oklahoma high school students fail to learn higher levels of science that lead to college graduation and scientific and technical careers. Particularly disadvantaged are Native Americans. According to the National Assessment of Educational Performance assessments from 1996 through 2003, Native American students nationally are scoring below other students at fourth, eighth, and twelfth grade levels. In addition, approximately 22% fewer Native American students complete the core coursework in high school.

Teachers in rural areas are often not prepared or certified in the subjects they are teaching. This is particularly true in science education where 24% of rural science teachers lacked academic majors or certification compared to 18% of teachers in nonrural settings (NSB, 2006). The K20 Center network of 500 rural schools provides an extensive infrastructure from which to design, implement, test, revise, and share results of rural education innovation. The K20 SCIENCE Initiative for Rural Schools described in this chapter directly impacts these low-income, rural schools. Although early in its implementation, participating rural teachers report strengthening of their own scientific understanding and how students develop scientific understanding; making complex changes in pedagogy to foster that development; and reinventing their practice in such a way as to reflect authentic intellectual work both in themselves and their students.

While the K20 SCIENCE Initiative directly impacts rural schools in the state of Oklahoma, the initiative has broad implications for schools nationally and perhaps internationally. CNN Polling Director Keating Holland examined U.S. Census data and identified 12 key statistics—four that measure race and ethnicity, four that examine income and education, and four that describe the typical neighbourhood in each state—and calculated how distant each was from the figures for the average state on each measure. Oklahoma ranked sixth nationally in the CNN poll of the most representative state in the country—the state that is a microcosm of the entire country (Preston, 2006). If Oklahoma is representative of the United States, the K20 SCIENCE Initiative has broader impacts than Oklahoma rural schools.

While equity in the United States is lacking for rural schools, other countries may or may not provide greater resources and support to their rural schools. Additionally, in some countries, the geographical setting at which inequity occurs may be different—for example, in some countries suburban and urban settings may be short-changed. The implication for governmental and educational leaders across the globe is that it is important to examine social justice and equity across geographical settings.

Although not every student will become a working scientist, all students, including rural students, need to make informed decisions as citizens about crucial science issues involving health, environment, energy, spending, and ethics (Conn, 2004). K20 SCIENCE advances social justice in rural schools through new conceptions of teacher professional development that enhances learning and prepares citizens for democratic participation.

NOTE

1. The K20 Center's programs, including K20 SCIENCE Initiative, are committed to the promotion of the democratic education **IDE-ALS—Inquiry, Discourse, Equity, Authenticity, Leadership, and Service,** and are grounded in the knowledge bases on school and community partnerships, teacher quality, and student success.

ACKNOWLEDGMENTS

The authors gratefully acknowledge support from Perri Applegate, Linda Atkinson, Jean Cate, Janis Slater, Leslie Williams, and Scott Wilson, researchers and staff of the K20 Center for Educational and Community Renewal, University of Oklahoma. This research was supported by the Oklahoma Medical Research Foundation through a Howard Hughes Medical Institute grant and the National Science Foundation's Research and Evaluation on Education in Science and Engineering (REESE) program (REC-0634070). Opinions reflect those of the authors and do not necessarily reflect those of the granting agencies.

REFERENCES

Atkinson, L. K. (2005). *Schools as learning organizations: Relationships between professional learning communities and technology-enriched learning environments.* Norman, OK: University of Oklahoma.

Atkinson, L. K., Williams, L. A., O'Hair, M. J., & O'Hair, H. D. (2008). Developing and sustaining schools as technology enriched learning organizations. *I-managers Journal of Educational Technology, 3*(4), 17 33.

Bass, H., Usiskin, Z. P., & Burrill, G. (Eds.). (2002). *Studying classroom teaching as a medium for professional development*. Proceedings of a U.S.-Japan workshop. Washington, DC: National Academy Press.

Beeson, E., & Strange, M. (2003). *Why rural matters 2003: The continuing need for every state to take action on rural education*. Randolph, VT: Rural School and Community Trust Policy Program.

Bishaw, A., & Stern, S. (2006). *Evaluation of poverty estimates: Comparison of the American Community Survey and the Current Population Survey*. Washington, DC: Poverty and Health Statistics Branch, U.S. Census Bureau. Retrieved August 15, 2008, from http://www.census.gov/hhes/www/poverty/acs_cpspovcompreport.pdf

Blumenfeld, P. C., Soloway, E., Marx, R. W., Krajcik, J. S., Guzdial, M., & Palincsar, A. (1991). Motivating project-based learning. *Educational Psychologist, 26*(3/4), 369–398.

Bransford, J. D., Brown, A. L. & Cocking, R. R. (Eds.). (2000). *How people learn: Brain, mind, experience, and school committee on developments of science of learning*. Washington DC: National Academies Press.

Bransford, J., Vye, N., Bateman, H., Brophy, S., & Roselli, B. (2004). Vanderbilt's AMIGO[3] project: Knowledge of how people learn enters cyberspace. In T. M. Duffy & J. R. Kirkley (Eds.), *Learner-centered theory and practice in distance education* (pp. 209–234). Mahwah, NJ: Erlbaum.

Carpenter, T. P. (2004, Spring). Scaling up innovative practices in math and science. *WCER Research Highlight, 16*(2), 1–8.

Conn, S. S. (2004, February 11). A new teaching paradigm in information systems education: An investigation and report on the origins, significance, and efficacy of the agile development movement. *Information Systems Education Journal, 2*(15). Retrieved June 22, 2005, from http://isedj.org/2/15/

Chokshi, S., & Fernandez, C. (2004). Challenges to importing Japanese lesson study: Concerns, misconceptions, and nuances. *Phi Delta Kappan, 85*(7), 520–525.

Darling-Hammond, L. (2000). Teacher quality and student achievement: A review of state policy evidence. *Education Policy Analysis Archives, 8*(1). Retrieved May 19, 2004, from http://epaa.asu.edu/epaa/v8n1/

Dewey, J. (1938). *Experience and education*. New York: Simon & Schuster.

Dufour, R., Eaker, R., & Dufour, R. (2002). *Getting started: Reculturing schools to become professional learning communities*. Bloomington, IN: National Educational Service.

Education Trust. (2003). *Education watch: The nation*. Retrieved May 8, 2006, from http://www2.edtrust.org/edtrust/summaries2004/USA.pdf

Fullan, M. (2001). *Leading in a culture of change*. San Francisco: Jossey-Bass.

Fullan, M. (2003). *Change forces with a vengeance*. New York: Routledge.

Garet, M. S., Porter, A. C., Desimore, L. M., Birman, B. T., & Yoon, K. S. (2001). What makes professional development effective? Results from a national sample of teachers. *American Educational Research Journal, 38*(4), 915–945.

Gates Foundation. (2004). *Southern governors committed to high performing rural schools*. Retrieved April 2, 2005, from http://www.gatesfoundation.org/ Education/Announcements/Announce-042104.htm

Good, T. L., & Brophy, J. E. (2000). *Looking in classrooms* (8th ed.). New York: Longman.

Greene, B. A., Miller, R. B., Crowson, H. M., Duke, B. L., & Akey, C. L. (2004). Relations among student perceptions of classroom structures, perceived ability, achievement goals, and cognitive engagement and achievement in high school language arts. *Contemporary Educational Psychology, 29*(4), 462–482.

Greeno, J. G. (1997). On claims that answer the wrong questions. *Educational Researcher, 26*(1), 5–17.

Hamlin, G. (2007). *An evaluation report to the Oklahoma Educational Technology Trust-concerning sustainability of Grants to Schools*. Oklahoma City, OK: OETT.

Hannafin, M. J., & Land, S. M. (1997). The foundations and assumptions of technology-enhanced, student-centered learning environments. *Instructional Science, 25*(3), 167–202.

Hord, S. M. (1997). *Professional learning communities: Communities of continuous inquiry and improvement*. Austin, TX: Southwest Educational Development Laboratory.

Huffman, J. B., & Hipp, K. K. (2003). *Reculturing schools as professional learning communities*. Lanham, MD: Scarecrow Education.

James, W. (1958). *Talk to teachers*. New York: W. W. Norton.

Jimerson, L. (2003). *The competitive disadvantage: Teacher compensation in rural America*. Washington, DC: Rural School and Community Trust.

Johnson, J., & Strange, M. (2007). *Why rural matters 2007: The realities of rural education growth*. Arlington, VA: Rural School and Community Trust.

Jonassen, D. H. (1999). Designing constructivist learning environments. In C. M. Reigeluth (Ed.), *Instructional design theories and models: A new paradigm of instructional technology* (Vol. 2, pp. 215–239). Mahwah, NJ: Erlbaum.

Kincheloe, J. L. (1991). *Teachers as researchers: Qualitative inquiry as a path to empowerment*. London: The Falmer Press.

Lamkin, M. L. (2006). Challenges and changes faced by rural superintendents. *Rural Educator*. Retrieved June 5, 2007, from http://findarticles.com/p/articles/ mi_qa4126/is_200610/ai_n16840815

Lee, V. E., & Smith, J. B. (1994, Fall). High school restructuring and student achievement: A new study finds strong links. *Issues in Restructuring Schools, 7*, 1–16.

Lee, V. E., & Smith, J. B. (1996). Collective responsibility for learning and its effects on gains in achievement for early secondary school students. *American Journal of Education, 104*(2), 103–147.

Lee, V. E., Croninger, R. G. & Smith, J. B. (1997). Course-taking, equity, and mathematics learning: Testing the constrained curriculum hypothesis in U.S. secondary schools. *Educational Evaluation and Policy Analysis, 19*(2), 99–121.

Lewis, C. (2002a). Does lesson study have a future in the United States? *Journal of the Nagoya University Education Department, 1*(1), 1–24.

Lewis, C. (2002b). *Lesson study: A handbook of teacher-led instructional change*. Philadelphia: Research for Better Schools.

Lewis, C., & Tsuchida, I. (1997). Planned educational change in Japan: The case of elementary science instruction. *Journal of Educational Policy, 12*(5), 313–331.

Lewis, C., Perry, R., & Hurd, J. (2004). A deeper look at lesson study. *Educational Leadership, 61*(5), 6–11.

Little, J. (1993). Teachers' professional development in a climate of educational reform. *Educational Evaluation and Policy Analysis 15*(2), 129–152.

Loucks-Horsley, S., Love, N., Stiles, K., Mundry, S. & Hewson, P. (2003). *Designing professional development for teachers of science and mathematics* (2nd ed.). Thousand Oaks, CA: Corwin.

Louis, K., Marks, H., & Kruse, S. (1996). Teachers' professional community in restructuring schools. *American Educational Research Journal, 33*(4), 757–798.

Malhoit, G., & Gottoni, N. (Eds.). (2003). The rural school funding report. *The Rural Education Finance Center, 2*, 12.

McCarty, R. (2003). Specified ignorance: A pedagogical and cognitive tool for learning the nature and process of science. *Teaching and Learning, 17*(3), 113–132.

National Center for Education Statistics. (2002). *The nation's report card: Mathematics 2000, NCES 2001–517*. Washington, DC: U.S. Department of Education, Office of Educational Research and Improvement.

Bass, H., Usiskin, Z., & Burrill, G. (2002). Studying classroom teaching as a medium for professional development. In *Proceedings of a U.S.-Japan workshop*. Washington, DC: National Academy Press.

National Science Board. (2006). *Science and Engineering Indicators 2006*. Washington, DC: National Science Foundation.

National Staff Development Council. (2001). *NSDC Standards*. Retrieved May 19, 2004, from http://www.nsdc.org/standards/index.cfm

Newmann, F. M. (1996). *Authentic achievement: Restructuring schools for intellectual quality*. San Francisco: Jossey-Bass.

Newmann, F. M. (2000, Fall). Authentic intellectual work: What and why? *Research/Practice 7*, 1. Retrieved March 16, 2004, from http://education.umn.edu/carei/Reports/Rpractice/Fall2000/default.html

Newmann, F. M., Bryk, A. S., & Nagaoka, J. K. (2001). *Authentic intellectual work and standardized tests: Conflict or coexistence?* Chicago: Consortium on Chicago School Research.

Newmann, F. M., King, M. B., & Youngs, P. (2000). Professional development that addresses school capacity: Lessons from urban elementary schools. *American Journal of Education, 108*(4), 259–299.

Newmann, F. M., & Wehlage, G. G. (1995). *Successful school restructuring: A report to the public and educators*. Madison, WI: Center on Organization and Restructuring of Schools, University of Wisconsin.

O'Hair, M. J., McLaughlin, H. J., & Reitzug, U.C. (2000). *Foundations of democratic education*. Fort Worth, TX: Harcourt Brace.

O'Hair, M. J., Williams, L. A., Wilson, S., & Applegate, P. (in press). Addressing social justice in rural schools: The K20 Model for systemic educational change and sustainability. In P. Woods & G. Woods (Eds.), *Leading alternative in education*. London: Palgrave Macmillan.

Pasley, J. D., Weiss, I. R., Shimkus, E., & Smith, P. S. (2004). Looking inside the classroom: Science teaching in the United States. *Science Educator, 13*(1), 1–11.

Piaget, J. (1972). *The psychology of the child*. New York: Basic Books.

Rovai, A. P. (2002). Building sense of community at a distance. *International Review of Research in Open and Distance Learning, 3*(1), Retrieved August 1, 2006, from http://www.irrodl.org/index.php/irrodl/article/view/79/152

Preston, M. (2006, July 27). The most "representative" state: Wisconsin. *Politics: The Morning Grind*. Retrieved August 22, 2006, from http://www.cnn.com/ 2006/POLITICS/07/27/mg.thu/index.html

Roeser, R. W., Midgley, C., & Urdan, T. (1996). Perceptions of the school psychological environment and early adolescents' self-appraisals and academic engagement: The mediating role of goals and belonging. *Journal of Educational Psychology, 88*(3), 408–422.

Roscigno, V. J., & Crowley, M. L. (2001). Rurality, institutional disadvantage, and achievement/attainment. *Rural Sociology, 66*(2), 268–293.

Senge, P. (2000). *Schools that learn: A fifth discipline field book for educators, parents, and everyone who cares about education*. New York: Doubleday.

Sergiovanni, T. J. (1994). *Building community in schools*. San Francisco: Jossey-Bass.

Silverman, P. (2003). The origins of molecular biology: A pedagogical tool for the professional development of pre-college science teachers. *Biochemical Molecular Biology Education, 31*(5), 313–318.

Slater, J., & Cate, J. M. (2006, April). *Cognitive dissonance as a perspective in the transfer of learning from authentic teacher research experiences to inquiry instruction in the classroom*. Paper presented at the American Educational Research Association Annual Conference, San Francisco.

Smith, J. B., Lee, V. E., & Newmann, F. M. (2001). *Instruction and achievement in Chicago elementary schools*. Chicago: Consortium on Chicago School Research.

Stewart, R. A., & Brendufer, J. L. (2005). Fusing lesson study and authentic achievement. *Phi Delta Kappan, 86*(9), 681.

Stigler, J. W., & Hiebert, J. (1999). *The teaching gap: Best ideas from the world's teachers for improving education in the classroom*. New York: Summit.

Tanner, K. D., Chatman, L., & Allen, D. (2003). Approaches to biology teaching and learning: Science teaching and learning across the school-university divide—Cultivating conversations through scientist-teacher partnerships. *Cell Biology Education, 2*(4), 195–201.

Vygotsky, L. S. (1978). *Mind and society: The development of higher mental processes*. Cambridge, MA: Harvard University Press.

Wellman, B. (1999). The network community: An introduction to networks in the global village. In B. Wellman (Ed.), *Networks in the global village* (pp. 1–48). Boulder: Westview Press.

Williams, L. A. (2006). *The relationships of technology integration and high school collaboration through the development of a professional learning community: A mixed methods study*. Unpublished doctoral dissertation, University of Oklahoma.

Williams, L., Atkinson, L., O'Hair, M. J., & Applegate, P. (2007, April). *Improving educational quality through technology-enriched learning communities for success in the global economy*. Paper presented at the meeting of the American Educational Research Association, Chicago.

Yoshida, M. (1999). *Lesson study: A case study of a Japanese approach to improving instruction through school-based teacher development*. Unpublished doctoral dissertation, University of Chicago.

READING 12

ASSESSING LEARNING FROM INQUIRY SCIENCE INSTRUCTION

By Stephanie B. Corliss and Marcia C. Linn

Measuring learning from science inquiry instruction has benefitted from the development of powerful technology-enhanced curriculum materials that feature scientific visualizations and embedded assessments. We review current assessment practices to measure students' higher order thinking in science and discuss promising approaches. We describe how the knowledge integration framework can inform the design of curriculum, assessments, and professional development to support complex scientific thinking. We conclude with recommendations for effective assessment practices and implications for future research.

RATIONALE AND RESEARCH QUESTIONS

Reforms in science education emphasize the importance of teaching and learning through scientific inquiry, where students engage in complex thinking as they construct their own understanding. Assessing this form of learning requires new measures aligned with instruction (American Association for the Advancement of Science [AAAS], 1993; National Research Council [NRC], 1996, 2001) and consistent with advances in understanding of student learning (Bransford, Brown, & Cocking, 2000). Assessments need to measure students' deep understanding of complex science topics and their ability to use the scientific inquiry processes in support of their learning.

Learning through inquiry requires students to utilize problem solving, analytical, and science process skills, as well as to understand the nature of science (Zachos, Hick, Doane, & Sargent 2000). Students have to develop research questions, form hypotheses, design investigations, analyze and interpret data, and form and evaluate scientific explanations (Kuhn, Black, Keselman, & Kaplan, 2000). These advanced skills are essential for functioning in an increasingly complex and scientifically sophisticated society (NRC, 1996, 2001) where one must think critically about often-persuasive messages and grapple with issues such as global climate change, energy conservation, and health decision making.

Implementing inquiry in the classroom can be challenging for both students and teachers. In an inquiry classroom students engage in scientific investigations while working together to solve open and complex problems. This student-centered approach puts the teacher into the role of facilitator rather than presenter of knowledge. Teachers must find ways to deal with students' multiple and sometimes conflicting ideas about science, and they need strategies for eliciting prior knowledge, adding new ideas, and assisting students in distinguishing between and reconciling their ideas about science in order to build a more coherent and cohesive understanding (Davis & Varma, 2008). Students

Stephanie B. Corliss and Marcia C. Linn, "Assessing Learning From Inquiry Science Instruction," *Assessment of Higher Order Thinking Skills*, ed. Gregory Schraw and Daniel H. Robinson, pp. 219-243. Copyright © 2011 by Information Age Publishing. Reprinted with permission.

are active in the learning process as they construct their knowledge. Research shows that inquiry instruction, combined with effective pedagogical practices and technology supports, can lead to improved student learning of science concepts (Krajcik et al., 1998; Linn, Davis, & Bell, 2004; Reiser et al., 2001; Sandholtz & Reilly, 2004; Scardamalia & Bereiter, 1996; Songer, Lee, & Kam, 2002). However, not all assessments are sensitive to inquiry instruction (Clark & Linn, 2003).

Traditional science assessments do not measure the type of complex thinking that is promoted in inquiry instruction; they mainly focus on recall of scientific facts (Lederman & Niess, 2000). More authentic assessments that emphasize complex thinking in science are needed to assess inquiry instruction (NRC, 1996, 2000; Pellegrino, Chudowsky, & Glaser, 2001). Furthermore, assessments need to align with innovative curricular materials and professional development opportunities to support inquiry teaching in the classroom (Linn, Lee, Tinker, Husic, & Chiu, 2006).

In this chapter, we discuss assessment of inquiry in the context of research projects that investigate technology-enhanced inquiry instruction and employ professional development that emphasizes inquiry learning.

TABLE 12.1. Lower Order and Higher Order Science Skills

SCIENCE SKILLS	LEARNING ACTIVITIES AND/OR ASSESSMENTS	
Lower order	Demonstrating knowledge of scientific concepts, laws, theory, procedures, and instruments	Recall
		Define
		Describe
		List
		Identify
Higher order	Applying scientific knowledge and procedures to solve complex problems	Formulate questions
		Hypothesize/Predict
		Design investigations
		Use models
		Compare/Contrast/Classify
		Analyze
		Find solutions
		Interpret
		Integrate/Synthesize
		Relate
		Evaluate

The assessment, curricular materials, and professional development are all aligned using the knowledge integration framework (Linn, 1995; Linn, Davis, Bell, 2004). This work draws on findings from the Technology-Enhanced Learning in Science (TELS; http://www.telscenter.org) Center for Teaching and Learning and the Mentored and Online Development of Educational Leaders in Science (MODELS; http://models.berkeley.edu) project. We consider the following research questions:

1. What are valid assessments of complex scientific thinking?
2. How can assessments capture students' ability to reuse complex thinking strategies in new settings?
3. How can assessment support teachers' as they promote complex thinking in science?

ASSESSMENT OF COMPLEX THINKING IN SCIENCE

Both national and international science assessment efforts recognize the importance of assessment of complex thinking in science. The NAEP (National Assessment of Educational Progress) science framework includes a section on practical reasoning, described as "the ability to apply appropriate scientific knowledge, problem-solving skills, and thinking processes to the solution of real problems." The assessment items are designed to probe students to think abstractly and consider hypothetical situations; consider several factors simultaneously; take an objective view of situations; and realize the importance of practical reasoning and experience (Grigg, Lauko, & Brockway, 2006). TIMSS (Trends in International Mathematics and Science Study) assessment framework also has a focus on advanced cognitive skills, such as knowing, applying and reasoning. Key skills of the reasoning domain include: analyzing problems to determine the relevant relationships, concepts, and problem-solving steps; making connections between concepts in different areas of science; combining knowledge of science concepts with information from experience or observation to formulate questions that can be answered by investigation; formulating hypotheses; making predictions; designing or planning investigations; making valid inferences based on evidence; applying conclusions to new situations; and constructing arguments to support the reasonableness of solutions to problems (Gonzales et al., 2008).

Efforts have been made to include open-ended items that measure complex thinking on these assessments, but these items are scored with a "right-or-wrong" or "right, partial, wrong" approach. These dichotomous scoring procedures fail to capture the nuances of students, thinking (Liu, Lee, Hofstetter, & Linn, 2008). Scoring methods with additional categories that emphasize links between relevant ideas about the scientific phenomena can provide more information about how students integrate their knowledge. We have investigated and refined the knowledge integration framework to design and score items that measure complex thinking in science.

KNOWLEDGE INTEGRATION FRAMEWORK

Knowledge integration focuses on how learners struggle with multiple and conflicting ideas in science, how they develop new ideas, and how they sort out connections between new and existing ideas to reach a more coherent understanding (Linn, 1995; Linn, Davis, & Bell, 2004). Students bring existing ideas to any learning situation. They come to science class with a repertoire of conflicting and often confusing ideas. These ideas evolve as students are presented with new ideas in science instruction. Students compare ideas, identify links and connections among their ideas, and gather evidence to support their ideas. Research shows that effective science instruction uses students' ideas as the starting point for investigating scientific phenomena and guides learners as they add new ideas, sort out these ides in a variety of contexts, make connections among ideas at multiple levels of analysis, develop a more nuanced criteria for evaluating ideas, and formulate linked views about the scientific phenomena (Linn, 2006; Linn et al., 2004).

The knowledge integration framework identifies four main categories of design principles to support meaningful learning: (1) making science accessible, (2) making thinking visible, (3) helping students learn from each other, and (4) promoting lifelong learning (Kali, 2006; Linn & Hsi, 2000; Linn, Clark, & Slotta, 2003). These principles guide the development of assessments to measure students' complex thinking in science, and also the design of curriculum, professional development, and instruction to promote complex thinking.

Designing Knowledge Integration Curriculum Materials

The curricular projects used in the research reported here were collaboratively designed by researchers and teachers at the TELS Center for Teaching and Learning. The projects guide students in scientific inquiry of real world issues through collaborative activities that emphasize investigation and the use of complex visualizations (Linn et al., 2006). Projects are used in 6th through 12th grade classrooms and cover topics in earth science, life science, physical science, biology, chemistry, and physics (see examples in Figures 12.1 and 12.2). Students and teachers access the projects through the Web-based Inquiry Science Environment (WISE; http://www.wise.berkeley.edu/). Each project takes approximately 7 to 10 fifty-minute science class periods to complete.

The activity structures in TELS projects are designed to engage students in the four interrelated processes of knowledge integration: eliciting ideas, introducing new ideas, developing criteria for evaluating ideas, and sorting and reorganizing ideas (Clark, Varma, McElhaney, & Chiu, 2008; Linn & Eylon, 2006). *Science is made accessible* in TELS projects by engaging students in inquiry investigations that connect to their everyday knowledge and reflect current scientific dilemmas such as interpreting claims about global warming, choosing appropriate treatments for cancer, or choosing an energy-efficient car. TELS projects *make thinking visible* by providing opportunities for students to interact with dynamic visualizations and to construct scientific explanations and visual

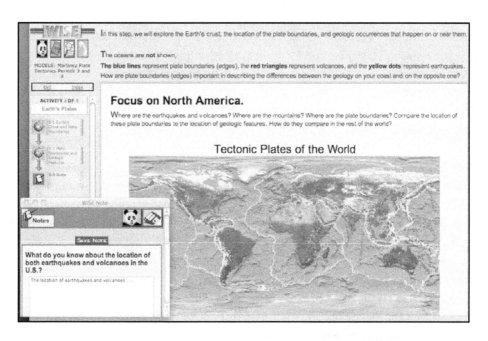

FIGURE 12.1: Screenshot of TELS project, *Plate Tectonics: What's on Your Plate?*, and embedded assessment item.

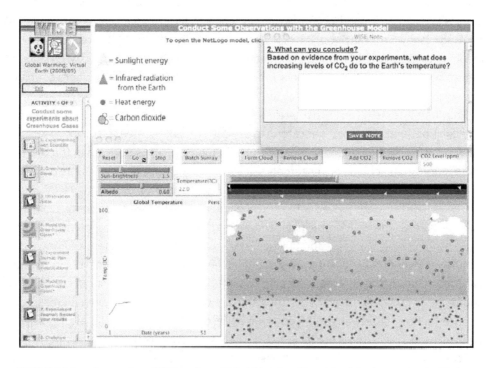

FIGURE 12.2: Screenshot of TELS project, *Global Warming: Virtual Earth Project*, and embedded assessment item.

models to represent their thinking. *Collaborative learning* is promoted throughout the projects with peer collaboration tools such as structured debates, online discussions, and share and critique activities. Finally, TELS projects encourage *lifelong science learning* by engaging students in reflecting on their own ideas and understanding in science, in critiquing diverse science information, and in developing inquiry skills useful for future science learning (Linn & Hsi, 2000).

Measuring Knowledge Integration

Knowledge integration assessments stimulate learning as well as evaluate learning. Like TELS curricular projects, the assessments engage students in complex thinking in various scientific contexts in everyday life. Items pose a dilemma and require students to generate an argument. Students must establish connections between new knowledge and existing knowledge, integrate knowledge gained from various sources, and warrant these ideas with evidence.

In TELS research studies, we use three types of assessments to evaluate how students integrate their knowledge of complex science topics: items embedded within TELS curricular projects; instruction sensitive pre/post tests, and annual benchmark tests including instruction sensitive items, standardized items, and questions about science learning. Items embedded within TELS curricular projects elicit students' ideas and explanations and encourage them to reflect on their understanding of the science concepts (see examples in Figures 12.1 and 12.2). This allows teachers to formatively evaluate students' thinking as they progress through the projects, which gives them opportunities to provide feedback to their students and to make changes to their teaching if necessary. Embedded assessments also provide researchers with information about how the project's cognitive and social supports contribute to students' learning.

Pre/posttests are given immediately preceding and following the enactment of TELS curricular projects. They are short four to six item tests that are instruction sensitive. They measure students' knowledge of the scientific concepts covered in the project. Pretests inform teachers and researchers of students' existing ideas while the posttests measure students' understanding at the end of the project.

Benchmark tests, completed at the end of each school year, serve as a delayed posttest that measures students' long-term understanding. These tests were designed to calibrate students' knowledge integration abilities, track students' development of knowledge integration over time, and compare student achievement with national norms on standardized science learning measures. TELS benchmark assessments include multiple choice items and explanation items. Students are often asked to explain their reasoning for their multiple-choice answer. Along with items developed by researchers, open-ended NAEP and TIMSS items are included in TELS assessments. Pre/posttest items and benchmark items are developed to measure students' learning of the science concepts both inside and outside of the context of the TELS projects.

All items are scored using a 5-point knowledge integration rubric that emphasizes students' links between relevant ideas about the scientific phenomena. Using a knowledge integration rubric, no answer is scored as a 0, and off task responses are scored as 1. Responses that show no links between normative ideas are scored as 2. A score of 3 represents a "partial link" where relevant ideas are expressed but not elaborated on to demonstrate how two ideas are connected. Responses that demonstrate a "full link" between two normative and relative ideas are scored a 4. Scores of 5 represent a "complex link" where two or more scientifically valid links are expressed. Figure 12.3 shows an example of an assessment item from the earth science benchmark test and the rubric used to score student responses.

A sophisticated study using the Rasch PCM item response theory (IRT) technique was conducted to gather reliability and validity data on the six TELS subject-specific (Earth Science, Life Science, Physical Science, Biology, Chemistry, Physics) benchmark tests (Liu et al., 2008). Over 3000 students took the TELS assessments at the end of the academic year. The number of items ranged from 16 to 27, depending on the test. Results provided using item response theory (IRT) provide satisfactory evidence for reliability and validity of knowledge integration assessments (Liu et al., 2008). They measure students' ability to sort, link, distinguish, and evaluate ideas and evidence. Explanation items are better at differentiating student science performance than multiple choice items, and using a knowledge integration scoring rubric on NAEP and TIMSS items greatly improves their psychometric properties. These assessments require students to think deeply about science content, which can promote an atmosphere of critical thinking in the classroom. In the following sections we report results from classroom studies where TELS curricular projects and knowledge integration assessments were used to engage both students and teachers in complex scientific thinking. In the following sections we report results from various classroom studies using knowledge integration assessments to measure complex scientific thinking. In the first study, students' knowledge integration of complex topics in earth science were examined after completing two TELS curricular projects. Next, we discuss the results of a study that assessed students' science concept knowledge integration and their ability to transfer general inquiry skills to novel tasks after completing a TELS curricular project. In the last study, we discuss the MODELS professional development program to facilitate teacher use of knowledge integration assessment and the effects on change in teacher practice. These studies combine TELS curricular projects and knowledge integration assessments to engage both students and teachers in complex scientific thinking.

Picture "A" shows a real greenhouse where light from the sun passes through the glass panels and heats the inside. The glass panels of the greenhouse keep the heat energy from escaping.

Picture A

Picture B

Picture "B" shows the greenhouse effect that happens on Earth. Which part of the picture is like the glass of the greenhouse? **Circle One.**

         SUN   SPACE   ATMOSPHERE   EARTH

Explain your answer.

Level	Criteria	Student Responses
No Answer		
Offtask		
Irrelevant or "I don't know."		
Irrelevant/Incorrect Have relevant ideas but fail to recognize links between them. Make links between relevant and irrelevant ideas. Have incorrect/irrelevant ideas.	-Scientifically incorrect statements, e.g., -Incorrect description of pictures, e.g., -Inaccurate connections between greenhouse and earth's atmosphere. -Says greenhouse is atmosphere, sun, earth, etc. without further explanation.	-It is sent back to the sun. -These pictures show about sun. -Arrows are pointing out -Atmosphere is like magnifying glass. -Ozone layer is like glass on the greenhouse.
Partial Have relevant ideas but do not fully elaborate links between them in a given context.	-Matches similar elements in two pictures but does not specify why/how they are similar. -Describe one picture as presented (usually focuses on how different arrows are pointing) -Mentions that atmosphere reflects, absorbs, or receives sunlight and does not relate one another.	-Because the sun reflects onto the plants and then goes back into space. -Because the earth suck some of the radiation into the earth lets it go slowly. -The dotted lines seem to show the glass and the dotted lines look like the atmosphere.

Basic Elaborate a scientifically valid link between two ideas relevant to a given context.	Mentions atmosphere is like the glass on a greenhouse and provides one of the following: -lets heat in and traps it there -keeping most of heat inside	- Earth is like the picture because the sunrays come from the sun heat up the Earth and sometimes reflect back.
Complex Elaborate two or more scientifically valid links among ideas relevant to a given context.	Mentions atmosphere is like the glass on a greenhouse and says "receive sunlight energy and blocks earth's (radiation) energy (or IR) from escaping to the space"	The glass is like the atmosphere. When solar energy hits the Earth, it turns into infrared energy (IR). If there are enough greenhouse gases the IR stays in Earth's atmosphere, heating up the Earth.

FIGURE 12.3: Earth science assessment item and knowledge integration rubric.

ASSESSING COMPLEX SCIENTIFIC THINKING IN EARTH SCIENCE

In the following study, we examined how two TELS curricular projects facilitate student knowledge integration of complex topics in earth science. We report results from knowledge integration assessments to evaluate the effectiveness of the curriculum and instruction and to demonstrate gains in students learning.

Participants and Procedures

Participants were 145 mixed-ability sixth grade science students from an ethnically and economically diverse school. All students were enrolled in a science class taught by the same teacher. The teacher had 33 years teaching experience. He taught science for 20 of the 33 years, and participated in the MODELS (Mentored and Online Development of Educational Leaders for Science) professional development program for 3 years. Students completed the *Global Warming: Virtual Earth and Plate Tectonics: What's on Your Plate?* projects during the 2007/2008 school year.

In both TELS projects, students interacted with complex scientific visualizations and then created their own model to demonstrate understanding of the scientific content. In *Global Warming*, students constructed a model of the Greenhouse Effect to show how global warming occurs and in *Plate Tectonics*, students created their own model to explain how a geological feature, such as a volcano or mountain, is formed. Students individually completed instruction sensitive pre/post tests for each project, and spent approximately 8 days working in pairs through each project during their science class time.

Data Sources and Analysis

The pre/posttests for each project consisted of four free response questions. Each item related to a main scientific concept to be learned in the unit (e.g., How does the Greenhouse Effect work to make sure that the temperature of the Earth's surface is just right?, List two causes you think contribute to global climate change. Why are most of the active volcanoes and earthquakes in the United States located on or near the West Coast?, What is the difference between a convergent and a collisional boundary?). Other questions required students to transfer their knowledge to a new context (e.g., According to the information in the table [surface temperature, composition of atmosphere], which planet has the greatest greenhouse effect? Explain your choice using evidence from the table., What evidence do scientists have for the movement of plates? How does this evidence support their ideas?). Responses to each item were coded using a 5-point knowledge integration rubric (see Figure 12.3).

Results

Results of paired t-tests revealed significant learning gains in students' complex thinking about the topics covered in the *Global Warming* ($t(125) = -11.04$, $p < .001$) and the *Plate Tectonics* ($t(130) = -10.44$, $p < .001$) projects. Table 12.2 shows students' learning gains. For *Global Warming*, mean composite scores increased from 4.71 on the pretest to 7.14 on the post-test. Mean composite scores increased from 5.70 on the pretest to 8.10 on the posttest for *Plate Tectonics*.

Results revealed that the TELS curricular projects were successful at helping students to develop a deep understanding of complex earth science concepts. Students demonstrated their ability to apply what they learned in the projects to both similar and novel questions on the posttest. Additionally, students were able to retain and apply their knowledge on end of year assessments; benchmark analyses revealed significant student learning gains in students' understanding of complex earth science concepts from the beginning to the end of the school year (Gerard, Spitulnik, & Linn, 2009). Results show that knowledge integration assessments measure students' ability to transfer their understanding of complex science concepts to contexts outside of the TELS curricular projects and that knowledge integration rubrics can capture progressively more sophisticated levels of students' reasoning in novel settings (Liu et al., 2008).

ASSESSING TRANSFER OF COMPLEX INQUIRY SKILLS

The goal of inquiry-based science instruction, like all education, is to promote the development of skills that students can transfer across disciplines and domains and use throughout their lifetime. The following study investigated how the guided inquiry experience presented in TELS projects facilitated the development of general inquiry skills that students could apply in a new learning context.

TABLE 12.2: Student Learning Gains From Pre- to Posttest

	PRETEST MEAN (SD)	POSTTEST MEAN (SD)	N
Global Warming	4.71 (5.41)	7.14 (2.08)	126
Plate Tectonics	5.70 (5.01)	8.10 (5.64)	131

To study transfer of general inquiry skills, researchers determined whether and when students could transfer the inquiry skills they engaged in during the TELS *Global Warming* project to a novel task (Corliss & Varma, 2008). Inquiry skills were defined as: (1) interpreting visual data, (2) using evidence to draw conclusions, (3) designing experiments, and (4) evaluating experimental results. These were the inquiry processes students must engage in to succeed in the TELS project. The transfer tasks required students to engage in these same processes in a context other than global warming.

Participants and Procedures

Fifty sixth grade science students at an ethnically diverse public school participated in this study. Students spent approximately 1.5 weeks working in pairs through the *Global Warming* project during their science class time. Within the project, students conducted experiments with a Greenhouse Effect visualization by manipulating levels of solar energy, atmospheric carbon dioxide, albedo, sunlight, and cloud cover. Following their investigations, students drew conclusions about the role of the different factors involved in the Greenhouse Effect. Students individually completed a 20-minute pretest and posttest measuring students' content knowledge and inquiry skills.

Data Sources and Analysis

Two items on the pre/posttests measured content knowledge (e.g., Describe how the Greenhouse Effect happens. What is the difference between the greenhouse effect and global warming?) and six items measured inquiry skills. The inquiry questions were designed to present students with visual data, allow students to draw conclusions based on evidence gathered from the visualization, have students design experiments to answer research questions, and evaluate conclusions based on experimental data (see Figure 12.4 for sample items). The inquiry questions differed from pre to posttest by context; students either answered questions about the survival of California fruit crops or California ocean species based on environmental factors. To account for any order effects, approximately half of the students received the fruit crops questions on the pretest and ocean species questions on the posttest, while the order was reversed for the remaining students. These items measured transfer of inquiry skills by requiring students to engage in the inquiry process outside of the context of the *Global Warming* project.

Two content and five inquiry questions from the pre/posttests were scored using 5-point knowledge integration rubrics. Separate composite scores were calculated for

15. Joe thinks that planting his citrus crops in the sun rather than in the shade will help his plants to produce more fruit. Describe how Joe can gather data to determine if his prediction is correct?

16. Amanda and George grow lemon trees. Their trees produced fewer lemons in 2006 than in previous years. They looked at data gathered the last 3 years to find out why. Their data is presented in the tables below.

Amanda's Data

	2004	2005	2006
Amount of Rainfall	62 inches	60 inches	36 inches
Fertilizer Use	Yes	Yes	Yes
Temperature Range	41°–90° F	40°–91° F	41°–91° F
Average # Lemons Produced	190	195	100

George's Data

	2004	2005	2006
Amount of Rainfall	62 inches	60 inches	36 inches
Fertilizer Use	Yes	Yes	No
Temperature Range	41°–90° F	40°–91° F	41°–91° F
Average # Lemons Produced	200	205	110

a. Which variable(s) differed in 2006 from previous years? Circle you answers)

  Rainfall   Fertilizer Use   Temperature Range   # of lemons

b. Amanda and George concluded that the lower amount of rainfall in 2006 than in previous years led to the decrease in the lemon fruit. Whose data best supports this conclusion? Circle your answer.

  Amanda's   George's   Both

c. Explain your answer using evidence from the data tables.

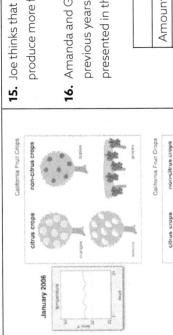

14. a. How do you think temperature affects California non-citrus crops?

b. Explain your answer using evidence from the diagram.

FIGURE 12.4: Items to measure students inquiry skills.

content knowledge and inquiry items. Table 12.3 provides details of the items included in the analyses and Table 12.4 contains a sample rubric.

Results

Results of a paired *t*-test revealed a significant difference between content items composite scores from pretest to posttest ($t(50) = -4.46$, $p < .01$) and inquiry items composite scores from pretest to posttest ($t(49) = -3.77$, $p < .01$.). Mean composite scores on the content items increased from 4.2 to 5.2, and inquiry item composite scores increased from 8.8 to 10.8. Students were able to transfer inquiry skills (e.g., interpreting visual data, using visual data as evidence to draw conclusions, designing experiments, and evaluating experimental results) to the novel tasks represented on the post-test. The *Global Warming* Project was effective in helping students to learn these skills as they were guided to manipulate the interactive Greenhouse Effect visualization and evaluate the results of their experiments to determine how particular variables affected global temperature.

Results of these studies revealed that as teachers and curriculum engage students in inquiry, students learn to integrate their ideas. Students not only gain content knowledge, but also valuable reasoning skills, both of which can be applied to novel contexts.

KNOWLEDGE INTEGRATION PROFESSIONAL DEVELOPMENT

The various types of knowledge integration assessments used in this research provide information about students' thinking before, during, and after their interactions with TELS projects. We have discussed how researchers use and interpret data gathered through knowledge integration assessments. However, the assessments were also designed to be used by teachers, but many may not know how to take advantage of these assessment opportunities to inform their instructional practices.

Professional development support can be a mediating factor to improve teachers' practices and enactment of inquiry-based science learning materials (Davis & Varma, 2008; Fishman et al., 2003; Schneider, Krajcik, & Blumefeld, 2005; Varma, Husic, & Linn, 2008). Professional development designed around analyzing student data reveal that teachers learn by improving their understanding of their own students' thinking (Fishman et al., 2003; Kazemi & Franke, 2004). TELS designed professional development to enable teachers to use knowledge integration inquiry assessments to improve their teaching.

TABLE 12.3: Pretest and Posttest Items Measuring Global Warming and Inquiry

ITEM	ITEM TYPE	SKILL MEASURED	SCORING RANGE
Q1	Explanation	Knowledge of Greenhouse Effect	0-5
Q2	Explanation	Knowledge of global warming	0-5

(*Continued*)

		Content Knowledge Composite Score (Q1+Q2)	0–10
Q13	Explanation	Interpreting visual data; Using evidence to draw conclusions	0–5
Q14	Explanation	Interpreting visual data; Using evidence to draw conclusions	0–5
Q15	Explanation	Designing experiment to gather data	0–5
Q16b	Multiple Choice	Evaluate experimental results	0–1
Q16c	Explanation	Evaluate experimental results	0–5
		Inquiry Skills Composite Score (Q13+Q14+Q15+Q16b+Q16c)	0–21

TABLE 12.4: Knowledge Integration Scoring Rubric for Experimentation Item

Q15: Joe thinks that planting his citrus crops in the sun rather than in the shade will help his plants to produce more fruit. Describe how Joe can gather data to determine if his prediction is correct?

SCORE	LEVEL	CRITERIA
0	No Answer	
1	Off Task	
2	Irrelevant/Incorrect	
3	Partial Have relevant ideas but do not fully elaborate links between them.	• Plant some crops in the shade and others in the sun (Independent Variable) and compare amount of fruit produced (Dependant Variable).
4	Full Elaborate a scientifically valid link between two relevant ideas.	• Plant some crops in the shade and others in the sun (Independent Variable) and compare amount of fruit produced (Dependant Variable). AND one of the following: • Discuss measurement technique of fruit crop and/or amount of crops planted • Discuss holding all other variables constant (water, fertilizer, soil type, ...)
5	Complex Elaborate two or more scientifically valid links among relevant ideas.	• Plant some crops in the shade and others in the sun (Independent Variable) and compare amount of fruit produced (Dependant Variable). AND both of the following: • Discuss measurement technique of fruit crops and/or amount of crops planted • Discuss holding all other variables constant (water, fertilizer, soil type, ...)

The knowledge integration perspective informed the design of professional development to support teachers in using technology-enhanced inquiry projects in their classrooms and aligning assessment practices with instruction. Teachers have a repertoire of ideas about science instruction, ideas about the content they teach, students' reasoning, pedagogical methods, and curriculum materials (Davis, 2004). Teachers acquire new ideas from many sources such as professional development opportunities, classroom experiences, teacher colleagues, or from examining their own students' ideas. When new ideas were encountered appropriate scaffolding can support teachers to reflect on their understanding, develop criteria to distinguish among ideas, and reorganize their knowledge to include the new ideas (Davis, 2004; Davis & Varma, 2008; Slotta, 2004).

We discuss professional development conducted by the MODELS program. MODELS is a 5-year National Science Foundation Teacher Professional Continuum project to support middle and high school science teachers as they plan, enact, and reflect on their experiences integrating TELS projects into their science curriculum. MODELS professional development activities are linked to the four principles of knowledge integration (Higgins, 2008). Activities are designed to make learning *accessible* and relevant to teachers by building on teacher' beliefs about science, technology, and their everyday practices. Students' *thinking is made visible* to teachers as they analyzed their students' work within the TELS projects and teachers' thinking is made visible as they responded to reflection questions following the professional development activities. MODELS provided teachers access to a supportive professional community, and activities were designed for teachers to *work collaboratively* so they could learn from each other. Professional development activities promoted *lifelong learning* by supporting teachers in continuously reflecting on their practice and planning modifications to their teaching.

MODELS PROFESSIONAL DEVELOPMENT

Approximately 20 middle school science teachers from two diverse school districts participated in the MODELS professional development. Teachers attended yearly summer professional development workshops and used at least one TELS project in the classroom each year. A goal of the professional development program was to support teachers in using evidence of students learning, measured by knowledge integration rubrics, to inform and refine their instruction and assessment practices. The professional development evolved over time as teachers' ideas about assessment become more sophisticated. In the following section, we discuss MODELS professional development activities and present evidence of changes in teachers' assessment practices and gains in their students' learning over time. A timeline of MODELS professional development and data collection is presented in Figure 12.5.

Workshop 1: Creating and Using Knowledge Integration Rubrics

At MODELS summer workshops, professional developers provided teachers with random samples of their students' worked from key embedded assessments within TELS projects. Teachers work collaboratively with others who used the same projects in their classrooms. At Workshop 1, teachers developed 5-point knowledge integration rubrics for the embedded assessment item and scored their student data. As students' thinking became visible to the teachers, they discovered gaps in their students understanding and made plans to address these challenging concepts the following year. Table 12.5 shows an example of one earth science teacher's data sheet from the professional development activity.

Following the professional development activity, one teacher commented, "It was very helpful to look through my students work using a very critical eye. I will need to consider if students need to have more direct instruction and class discussion to influence better comprehension of challenging information." Teachers used this evidence of student understanding to base changes to their teaching methods, assessment practices, and the technology projects.

Workshop 2: Formative Assessment and Feedback for Knowledge Integration

During Year 1, teachers reported using knowledge integration rubrics to formatively evaluate students' thinking, but complained of little time to provide individual feedback to each student about the quality of all of their work within TELS projects. The online environment allowed teachers to respond to students work by typing in specific comments or inserting premade comments that were viewable to the students. The assessment tool retained a list of these premade comments that could be used repeatedly by the teachers with multiple students or over multiple years using the TELS projects. Professional development activities that emphasized more efficient use of the assessment tool to provide effective feedback to students were implemented at Workshop 2.

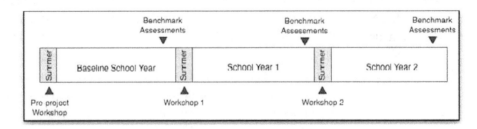

FIGURE 12.5: MODELS professional development and data collection timeline.

TABLE 12.5: Teacher Data Sheet From Summer Workshop 1

Project Title: Plate Tectonics: What's On Your Plate?
Activity: 3 Step: 2
Question: What do you know about the location of both earthquakes and volcanoes in the U. S.?

CATEGORIES OF STUDENT RESPONSES	EXAMPLES OF STUDENT RESPONSES	NUMBER OF STUDENTS IN GROUP PERCENTAGE
1. Off-task Students give no response, or give a response that does not answer the question.	The location of earthquakes are in the United States and the location of volcanoes are in the United States. (re-stated question = no answer)	1%
2. Irrelevant Link Student has misconceptions or makes invalid connections between ideas.	The location of earthquakes happen along the coast lines. The locations of volcanoes happen where the coastlines meet. The location of earthquakes are in flat areas and the location of volcanoes are in mountains.	14%
3. Partial Link Student needs to elaborate, or student's response is insufficient to solve the problem.	The location of earthquakes is where the plates meet. The location of volcanoes is where the magma pushes out of the ground. The locations of the earthquakes are on the west side of the U.S. The location of volcanoes are on the West Coast of the United States.	39%
4. Single Link Student makes one complete, correct connection.	Locations of earthquakes is on or near plate boundaries. The location of volcanoes is near plate boundaries. The location of earthquakes is normally on a plate boundary or fault line. The location of volcanoes is usually on a plate boundaries.	42%
5. Multiple Links Student makes two or more correct connections.	The location of earthquakes and volcanoes is at the plate boundaries. There is a plate boundary on the West Coast and that's why they are both there.	4%

Teachers collaboratively scored random samples of their students work again and compared the results to those from the previous year. The teachers then reviewed their projects and selected one or two key questions they would grade during project enactments the following year. They created knowledge integration rubrics for these specific questions. As they created the rubrics, they also created generic comments that related to each knowledge integration category and would be applicable to students' work in multiple TELS projects. Table 12.6 shows examples of comments created during the workshop activity. Individual teachers then entered the comments they planned to use during the upcoming school year into the TELS teacher portal.

Following the workshop activity, one teacher commented, "Working on the comments was extremely helpful. They are generic enough to work for all projects, yet specific enough to encourage 'better' answers. I hope that overall I will see more evidence of understanding." Teachers felt the professional development activity was worthwhile and were excited about using the new assessment technique in the upcoming school year to support students thinking about the science concepts.

TABLE 12.6: Examples of Teacher Comments Created During Professional Development Activity at Workshop

TEACHER COMMENTS
Needs work: Reread the question/prompt. This step needs corrections! Check for accuracy of writing/data. Try again!
Wrong-Redo: Your information is incorrect. Redo this step using better information/data. Do more research/discussion with your partner.
Evidence: You need to state facts or information based on reading or experiment data. Be specific with examples and/or real world connections.
Wow: This is great! You're providing complete answers using scientific vocabulary and evidence!
Partial link: Good start! Add more information or connections.
Evidence: Good answer. Now include specific evidence to support your statements.

Changes in Teachers' Assessment Practices

To examine how teachers' assessment practices changed after engaging in professional development, researchers reviewed records of all electronic comments given to their students through the teacher assessment tool during Year 1 and Year 2. Six teachers, one from each grade level at each school, were chosen for analysis. Each teacher attended both Workshop 1 and Workshop 2, and enacted one or more TELS projects during Year 1 and Year 2. We found that all 6 teachers graded more student work and provided more comments to their students in Year 2. The total comments given by all 6 teachers increased from 2,220 to 4,641 (see Table 12.7).

Results of the MODELS project revealed that by aligning curriculum, professional development, and assessment we saw changes in teachers' assessment practices to support students' higher order thinking. Professional development that engages teachers in the process of creating and using knowledge integration rubrics provided insight into students' thinking and evidence to base changes to the teaching and assessment practices. Further professional development on formative assessment techniques with the TELS teacher tools led to more efficient and effect strategies for examining students' thinking and providing feedback to students in the classroom. The result is better instructional practices and advances in students' higher order thinking about the science concepts presented in TELS projects over time (Gerard, Spitulnik, & Linn, 2009).

DISCUSSION

Valid assessments of inquiry learning in science are emerging in several research programs. These assessments ask students to make links among their ideas and to buttress their conclusions with evidence, and require students to apply their knowledge and skills to novel contexts.

TABLE 12.7: Teacher Electronic Feedback to Students in TELS Projects

SCHOOL, GRADE LEVEL	TELS PROJECT	NUMBER OF COMMENTS YEAR 1	NUMBER OF COMMENTS YEAR 2
Midvale, 8th	Velocity	149	319
Alta Vista, 8th	Velocity	260	346
Midvale, 7th	Mitosis	298	1367
Alta Vista, 7th	Mitosis	857	1371
Midvale, 6th	Plate Tectonics	422	472
	Global Warming	21	278
Alta Vista, 6th	Plate Tectonics	213	488
Total Comments		2220	4641

These innovative assessments need to align with inquiry instruction and professional development. Such alignment can change the interactions among students, teachers, and the curriculum. For schools to successfully integrate science inquiry learning, these elements need to change together. The research reported here provides evidence for the value of aligning curriculum, professional development, and assessment using the knowledge integration framework. When these practices support students to think more critically about complex science topics, we see improvement in student outcomes (Gerard et al., 2009).

Design of valid assessment requires more complex items than are typically used in state and national assessments. Assessment needs to move away from mostly multiple-choice items that do not make students thinking visible or measure students' deep understanding of scientific reasoning. Our research reveals that assessments designed for knowledge integration can accurately measure complex thinking in science inquiry. Multiple-choice assessments alone are not sensitive enough to measure deep understanding in science. Open-ended explanation items have the potential to tap into complex thinking as long as they are scored with an effective rubric that distinguishes between responses. For example, the 5-point knowledge integration rubrics successfully distinguish among students and are sensitive to inquiry teaching. Knowledge integration assessments have been used successfully to evaluate the effectiveness of TELS projects, to examine how students reuse complex thinking skills in new contexts, and to engage teachers in analyses of their students' thinking to inform changes to instructional and assessment practices.

Innovative assessments that measure inquiry learning can help teachers learn new practices by validly measuring student progress. Effective assessments help teachers identify and deal with students' multiple and sometimes conflicting ideas about science. Such assessments also encourage teachers to use the knowledge integration instructional pattern by revealing the multiple ideas that students hold and the limits of their ability to make links among ideas. The pattern calls for eliciting the prior knowledge of students in order to guide students to make links between their existing and new ideas. Second, the pattern calls for adding new, normative ideas that allow students to examine their own ideas. Third, the pattern calls for assisting students in distinguishing between the new ideas and their own views. When students develop scientific criteria for distinguishing ideas they are also able to apply these ideas to new situations. Fourth, the pattern calls for helping students sort out the repertoire of ideas that they hold and encouraging them to build a more coherent and cohesive understanding.

Professional development activities that engage teachers in using and creating knowledge integration rubrics give them better insight into their students' understanding and contribute to effective inquiry instruction. By analyzing student responses, teachers gain a deeper understanding of how students struggle to understand complex topics. Insights into student responses can help teachers improve their own assessment practices.

The studies reported here illustrate how assessment can shape learning and instruction. First, by giving knowledge integration pretests, the inquiry curriculum helps students understand the goals of inquiry instruction. Second, by embedding additional knowledge integration prompts in the instruction, the inquiry projects reinforce the importance of knowledge integration. Third, by giving teachers the opportunity to score student responses we contribute to teacher learning about the nature of inquiry and enable them to set realistic goals for their students. This activity can motivate teachers to consider ways to improve their own instructional strategies. Fourth, by using knowledge integration assessments to assign student grades, teachers elevate inquiry to a major goal of the course. Finally, by asking students to generalize their reasoning to novel problems, the assessments communicate that inquiry skills have multiple uses.

Future Research

The findings reported here show the benefit of aligning professional development, curriculum, and assessment to improve inquiry outcomes using the knowledge integration framework. We need a more nuanced understanding of how each of these aspects of science instruction contributes to student learning.

The value of knowledge integration assessments is clear but many questions remain. We know that knowledge integration measures are more sensitive than traditional multiple-choice outcome measures, but they are also more costly to score. In addition, teachers benefit from continuous information about student progress that can be gathered from embedded assessment. We need to understand how assessments can be used as learning events to improve teaching and learning over time. Research could help identify ways to design embedded assessments to replace current standardized tests and provide more informative and timely information for teachers.

We have glimmers of insight into how to promote and measure transfer of inquiry skills to new learning situations. We need more systematic and detailed studies of this question. Finally, we need more information about the long-term impact of inquiry versus traditional teaching. Additionally, as technology-enhanced instruction becomes more sophisticated, unique assessment opportunities will become available for teachers and researchers.

ACKNOWLEDGMENTS

This work was conducted in collaboration with others at the Center for Technology Enhanced Learning in Science (TELS) and researchers involved in the Mentored and Online Development of Educational Leaders in Science (MODELS) project, both of which are supported by the National Science Foundation under grant numbers 0334199 and 0455877. Any opinions, findings, and conclusions or recommendations expressed in this material are those of the authors and do not necessarily reflect the views of the National Science Foundation.

The authors wish to thank the members of the Technology-Enhanced Learning in Science (TELS) center, the Mentored and Online Development of Educational Leaders in Science (MODELS) project, and the teachers and students who participated in this work. Special thanks go to Michele Spitunik, the leader of the MODELS project, and Hee Sun Lee, the Director of Assessment for TELS.

REFERENCES

American Association for the Advancement of Science. (1993). *Benchmarks for science literacy*. New York, NY: Oxford University Press.

Bransford, J. D., Brown, A. L., & Cocking, R. R. (Eds.). (2000). *How people learn: Brain, mind, experience, and school*. Committee on Developments in the Science of Learning. National Research Council. Washington, DC: National Academy Press.

Clark, D., & Linn, M. C. (2003). Designing for knowledge integration: The impact of instructional time. *The Journal of the Learning Sciences, 12*(4), 451–494.

Clark, D. B., Varma, K., McElhaney, K., & Chiu, J. (2008). Design rationale within TELS projects to support knowledge integration. In D. H. Robinson & G. Schraw (Eds.), *Recent Innovations in educational technology that facilitate student learning* (pp. 157–193). Charlotte, NC: Information Age Publishing.

Corliss, S. B., & Varma, K. (2008, March). *Supporting the Development of Inquiry Skills in Technology-enhanced Science Curricula*. Paper presented at the annual meeting of the American Educational Research Association, New York, NY.

Davis, E. A. (2004). Knowledge integration in science teaching: Analyzing teachers' knowledge development. *Research in Science Education, 34*(1), 21–53.

Davis, E. A., & Varma, K. (2008) Supporting teachers in productive adaptation. In Y. Kali, M. C. Linn, & J. E. Roseman (Eds.), *Designing coherent science education*. New York, NY: Teachers College Press.

Fishman, B. J., Marx, R. W., Best, S., & Tal, R. T. (2003). Linking teacher and student learning to improve professional development in systemic reform. *Teaching and Teacher Education, 19*, 643–658.

Gerard, L. F, Spitulnik, M., & Linn, M. C. (2010). Teacher use of evidence to customize inquiry science instruction. *Journal of Research in Science Teaching, 47*(9), 1037–1063.

Gonzales, P., Williams, T., Jocelyn, L., Roey, S., Kastberg, D., & Brenwald, S. (2008). *Highlights From TIMSS 2007: Mathematics and Science Achievement of U.S. Fourth- and Eighth-Grade Students in an International Context* (NCES 2009–001). National Center for Education Statistics, Institute of Education Sciences, U.S. Department of Education. Washington, DC.

Grigg, W. S., Lauko, M. A., & Brockway, D. M. (2006). *The Nation's Report Card: Science 2005* (NCES 2006–466). Washington, DC: U.S. Department of Education, National Center for Education Statistics.

Higgins, T. E. (2008). *Through the eyes of professional developers: Understanding the design of learning experiences for science teachers*. Unpublished doctoral dissertation, University of California, Berkeley.

Kali, Y. (2006). Collaborative knowledge building using the Design Principles Database. *International Journal of Computer Support for Collaborative Learning, 1*(2), 187–201.

Kazemi, E., & Franke, M. L. (2004). Teacher learning in mathematics: using student work to promote collective inquiry. *Journal of Mathematics Teacher Education, 7*(3), 203–235.

Krajcik, J. S., Blumenfeld, P. C., Marx, R. W., Bass, K. M., Fredricks, J., & Soloway, E. (1998). Inquiry in project-based science classrooms: Initial attempts by middle school students. *Journal of the Learning Sciences, 7*(3 and 4), 313–350.

Kuhn, D., Black, J., Keselman, A., & Kaplan, D. (2000). The development of cognitive skills to support inquiry learning. *Cognition and Instruction, 18*(4), 495–523.

Lederman, N. G., & Niess, M. L. (2000). Problem solving and solving problems: Inquiry about inquiry. *School Science and Mathematics, 100*(3). 113–116.

Linn, M. C. (1995). Designing computer learning environments for engineering and computer science: The scaffolded knowledge integration framework. *Journal of Science Education and Technology, 4*, 103–126.

Linn, M. C. (2006). The knowledge integration perspective on learning and instruction. In R. K. Sawyer (Ed.), *The Cambridge handbook of the learning sciences* (pp.243–264). New York, NY: Cambridge University Press.

Linn, M. C., Clark, D., & Slotta, J. D. (2003). WISE design for knowledge integration. *Science Education, 87*(4), 517–538.

Linn M. C., Davis E. A. & Bell P. (2004) Inquiry and technology. In M.C. Linn, E. A. Davis & P. Bell (Eds.), *Internet environments for science education* (pp. 3–28). Mahwah, NJ: Erlbaum.

Linn, M. C., & Eylon, B. -S. (2006). Science Education: Integrating Views of Learning and Instruction. In P. A. Alexander & P. H. Winne (Eds.), *Handbook of Educational Psychology* (2nd ed., pp. 511–544). Mahwah, NJ: Erlbaum.

Linn, M. C., & Hsi, S. (2000). *Computers, teachers, peers: Science learning partners.* Mahwah, NJ: Erlbaum.

Linn, M. C., Lee, H. S., Tinker, R., Husic, F., & Chiu, J. L. (2006). Teaching and assessing knowledge integration in science. *Science, 313*, 1049–1050.

Liu, O. L., Lee, H.S., Hofstetter, C., & Linn, M. C. (2008). Assessing knowledge integration in science: Construct, measures, and evidence. *Educational Assessment, 13*, 33–55.

National Research Council. (1996). *The National Science Education Standards.* Washington, DC: National Academy Press.

National Research Council. (2001). *Inquiry and the National Science Education Standards.* Washington, DC: National Academy Press.

Pellegrino, J. W., Chudowsky, N., & Glaser, R.(Eds.).(2001). *Knowing what students know: The science and design of educational assessment.* Washington, DC: National Research Council.

Reiser, B. J., Tabak, I., Sandoval, W. A., Smith, B. K., Steinmuller, F., & Leone, A. J. (2001). BUGILE: Strategic and conceptual scaffolds for scientific inquiry in biology classrooms. In S. Carver & D. Klahr (Eds.), *Cognition and instruction: Twenty-five years of progress* (pp. 263–305). Mahwah, NJ: Erlbaum.

Sandholtz, J. H., & Reilly, B. (2004). Teachers, not technicians: Rethinking technical expectations for teachers. *Teachers College Record, 106*(3), 487–512.

Scardamalia, M., & Bereiter, C. (1996). Adaptation and understanding: A case for new cultures of schooling. In S. Vosniadou, E. DeCorte, R. Glaser, & H. Mandl (Eds.), International perspectives on the design of technology-supported learning *environments* (pp. 149–163). Mahwah, NJ: Erlbaum.

Schneider, R. M., Krajcik, J., & Blumenfeld, P. (2005). Enacting reform-based science materials: The range of enactments in reform classrooms. *Journal of Research in Science Teaching, 42*(3), 283–312.

Slotta, J. D (2004). The Web-based Inquiry Science Environment (WISE): Scaffolding teachers to adopt inquiry and technology. In M. C. Linn, P. Bell, & E. Davis (Eds.), *Internet Environments for Science Education* (pp. 203–232). Mahwah, NJ: Erlbaum.

Songer, N. B., Lee, H. S., & Kam, R. (2002). Technology-rich inquiry science in urban classrooms: What are the barriers to inquiry pedagogy? *Journal of Research in Science Teaching, 39*(2), 128–150.

Zachos, P. A., Hick, T. L., Doane, W. E. J., & Sargent, C. (2000). Setting theoretical and empirical foundations for assessing scientific inquiry and discovery in educational programs. *Journal of Research in Science Teaching, 37*(9), 938–962.

READING 13

MORE BASIC SCIENCE PROCESS SKILLS

MEASURE, PREDICT, AND COMMUNICATE

By John Settlage and Sherry Southerland

HIGHLIGHTS

- Students use tools to measure and to extend their observing. Measurements numerically express observations of objects and events.

- As students become more experienced with measuring, their proficiency improves. Giving students time to use rulers, balances, and other measuring tools will help them become comfortable with the metric system.

- The human desire to overcome uncertainty makes predicting a powerful skill. Predictions are powerful, however, when they are correct or incorrect. When students make incorrect predictions, their improper assumptions become explicit and identify where more thinking is required.

- Students communicate throughout the school day. When communicating is connected to their use of observing, inferring, or any of the other process skills, the communication becomes a science process skill.

- Activities that do not involve at least some of the process skills do not qualify to be called science.

- A joint productive activity in which students and teachers participate together in science activities represents an effective pedagogy for diverse learners.

- Viewing the world from a science perspective is an important ability. However, there are other ways of perceiving our surroundings that are legitimate ways of knowing even though they are not called scientific.

The science process skills describe the actions within the culture of science that students can develop through practice. Previously we considered three basic science process skills: observing, inferring, and classifying. In this chapter, we will examine the process skills of measuring, predicting, and communicating. As in the previous chapter, we will define each process skill, demonstrate how each is a component of the culture of science, and discuss where teachers need to pay special attention to students with differing cultures, home languages, or cognitive abilities.

John Settlage and Sherry Southerland, "More Basic Science Process Skills: Measure, Predict, and Communicate," *Teaching Science to Every Child: Using Culture As a Starting Point*, pp. 83-104. Copyright © 2012 by Taylor & Francis Group. Reprinted with permission.

ACTIONS OF SCIENCE AS ESSENTIAL

It is a common misperception that process skills are somehow less important than the science content we expect students to learn. Our contention is that separating content from process, or disconnecting the objects of science from the actions of science, is not a good idea. First, nothing will more quickly destroy students' natural interest in science than teaching only information without allowing them to actively participate in it. Second, developmental psychologists have demonstrated that elementary school students think in concrete ways; if science is taught in the abstract, it's not going to be useful to the students. Third, even though our attention is often drawn to the concepts when we consult state or national standards for science, the process skills are present in some form within these documents. Fourth, to teach content without adequate attention to process skills is an incomplete and erroneous representation of science as it is performed by professional scientists. Were we to teach without developing students' abilities to use the process skills, we would be teaching not science but actually some other odd subject that has little relationship to the culture of science. Finally, when we consider teaching science to students with diverse cultures, languages, and cognitive abilities, we must develop shared experiences, language skills, and conceptions. In summary, including process skills within science instruction will assist us in our goal of making science accessible to all learners.

MEASURING

As is true for all six of the basic science process skills, measuring is a skill that students need to perform and not just learn about. *Learning to measure* takes on greater significance and serves a higher instructional purpose than *learning about measurements*. Measuring is a special example of observing. When we observe we rely on our senses to gather information about the environment. Observations can be divided into two groups: if a number is somehow used to describe what is noticed, then we can label those observations as **quantitative**. In contrast, observations that do not make use of numbers are **qualitative**. To keep these two ideas straight, notice that the first term hints at the quantity of something whereas the second term suggests the quality of what's being observed.

It would be inappropriate to suggest that quantitative observations are superior to qualitative observations. Both are very important and have a place within science, because each way of observing has its own strengths. To describe the intensity of sunlight, we might use a light-sensing tool to quantify the brightness. When we are making observations of the moon, we describe its shape—which represents a qualitative observation.

The process skill of **measuring** is a special type of quantitative observation. This is because measuring requires the use of some tool, such as a ruler, to assist our observing. There are quantitative observations that contain numbers that are not measurements, because those observations aren't compared to a standardized measuring tool. Counting is the best example of this. In observing this page of text, you might count the number of times the letter "e" appears. This would not require you to use a measuring tool such as a ruler. Counting the hundred or more times that letter appears on this page is an example of quantitative observing but not an example of measuring.

Extending Your Senses

Because measuring involves using a tool, it is a special form of observing quantitatively. The tool is a standard against which objects can be compared. Although you might be able to look at a crayon and a pencil and visually determine which one is longer, a ruler can assist the observing by refining what you see. You can think of the ruler as a measuring tool that extends your sense of sight. When people are measuring, and this holds true for professional scientists and students, they use a tool that has been calibrated with a standard unit of measure. As a result, it makes the comparison of observations more precise. Instead of saying that a leaf is big, we can use a ruler to find the length and width of the leaf. If someone else observes the same leaf, they may not agree that it is big. If they measure the leaf, their observation should be the same.

Our senses are very useful to us, but they can be fooled. Tools that measure are not so easily deceived. Think about the cylinders you see in everyday life: a soup can, a paint can, a long potato chip can, a tennis ball can, and a fresh piece of chalk. Consider which would be the closest to having the same height as its circumference. Using your sense of sight you could compare the cylinders' height and circumference. It's hard to tell for sure. However, you can extend your sense of sight by relying on a measuring tape to compare the height and circumference to "see" what the right answer is.

Science Tools for Special Needs Students

It would be very natural for us to think about ourselves if we were learning science where our teacher was using the process skills. But as we develop our sensitivities to diverse populations, we will recognize the value in thinking about how those who are not just like us would respond. For example we might appreciate that learning to use a graduated cylinder would be a challenge for an elementary school student. In addition, we should consider the additional challenges a child who has a physical or visual disability might face when learning such a skill. The act of pouring a liquid into a narrow container and then using one's eyes to decide how many milliliters of liquid are present is a feat for anyone. But we are advocating for all students to participate in science. How might it be possible for students with physical limitations to perform something like the action just described?

We are fortunate that the staff at the Lawrence Hall of Science in California has created science tools to address these needs. The special but inexpensive pieces of equipment are part of a curriculum program called Science Activities for the Visually Impaired/Science Enrichment for Learners with Physical Handicaps. The shorter name is usually much easier to verbalize: SAVI/ SELPH (pronounced as "savvy self"). Figure 13.1 shows pictures and descriptions of some of this equipment from the SAVI/SELPH website.

We are not suggesting that as a new teacher you must obtain this specialized equipment or else visually impaired students will not be able to participate in science. But these images show that such materials are available, and you as a teacher may be able to access funds for purchasing such materials for special needs students. Beyond that, we hope this illustrates that with ingenuity and forethought we can engage all students in science. A physical impairment is not a sufficient cause to exclude children from the excitement of doing legitimate school science.

Reducing Sources of Bias

What we notice about our surroundings is influenced by our previous experiences. A toddler might point to the moon and notice how brightly it shines. Someone with more experience might notice its phase and position in the sky. The different backgrounds people have will cause them to note different features. Another example: people can be seated in the same room, and some will complain how cold it is whereas others are quite comfortable. The difference is because of the temperatures the various people are accustomed to and find comfortable. Again, each individual perceives the environment in his or her own way.

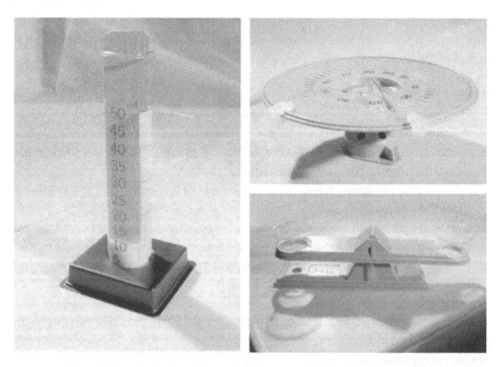

FIGURE 13.1: This equipment allows students with visual impairments to measure volume, mass, and temperature. (Photos courtesy of Lawrence Hall of Science at the University of California, Berkeley.)

Our use of measurements can help us avoid having our biases interfere with our observations. You can use a thermometer to augment your sense of touch and find the temperature of the room in terms of a standard unit of measure. Whether 20°C is comfortable is a personal preference, but this measurement will be the same no matter who does the measuring. Thermometers and other measuring tools are immune to the opinions of others. You may be likely to agree with grandmother when she asks, "Isn't it cold in here?" but measuring tools are less prone to the powers of suggestion.

Developing Measuring Proficiency

To become proficient with measuring, we have to develop several abilities. First, we need to learn how to properly use the measuring tool. For example when using a ruler to measure length, you have to check to make sure that the zero part of the ruler is at the end of the object you're measuring. You cannot just slap down the ruler and read the number. Before finding the mass of an object, you need to first make sure that the balance or scale reads zero when nothing is on it. One part of developing measuring skills is knowing the proper way to use the measuring tools.

Related to knowing how to use measuring tools properly is recognizing the proper tool to use to make the measurement. You compare heaviness with certain tools, and you compare height with another set of tools. Knowing that different tools provide different information becomes relevant when there is a question about which thing is bigger: what tool you use to compare pumpkins or apples depends on what type of measurement you're interested in studying. If "bigger" is to be decided by heaviness, then a scale or balance is needed. If "bigger" implies the girth, then something like a measuring tape should be used.

FOR REFLECTION AND DISCUSSION

Measuring requires that we use a tool to extend our senses and assign a quantitative value to the observations. A ruler extends our sense of sight, but measuring tools for the other senses may be less obvious. What are tools that can be used by scientists (or in everyday life) to extend the other senses of taste, touch, hear, and smell?

In addition to knowing the ways of measuring, you also should recognize which tool is most appropriate for making those kinds of measurements. A bathroom scale is not going to help determine which apple is the heaviest. A pocket ruler is not the best tool to use to measure the length of the hallway. A diet scale may not be appropriate for comparing the heaviness of shoes, and a stopwatch is not the right tool for measuring the amount of time that elapses between one full moon and the next. One way to help students choose the best tool for a particular measurement is to provide them with many opportunities to use a variety of measuring tools. Sometimes the teacher may provide some advice and guidance, but there may well be times to allow students to make their own measuring decisions. After students have a few instances of making less than useful measurements, the discussion of the most appropriate tool takes on a whole new significance. The choice of measuring tool becomes more than a matter of being told by the teacher—the students know from their experiences when to select one tool or another.

Metrics in the Science Classroom

What about the metric system? What place does it have in the school science curriculum? Here are two answers to these questions. The first is philosophical: the metric system is the way that scientists around the world make their measurements. Even scientists who collaborate only with other scientists in the United States use grams to record mass, meters to record length, and degrees Celsius to record temperature. This suggests that having students learn the metric system should be part of becoming scientifically literate. For students who have immigrated to the United States from other countries, our use of inches and pounds instead of centimeters and kilograms is going to seem odd. A teacher who is reluctant to use the metric system might need to overcome that reluctance to support the learning of students for whom the metric system is familiar and comfortable.

But the second response to the question about using the metric system is much more practical and depends on the local science standards. Are the students expected to understand centimeters when they take their standardized tests? If they are, then teaching the metric system to them is your obligation. Are the students supposed to know the metric system when they reach later grade levels? If high school science teachers expect the students to walk in the door with a familiarity with grams and meters, then, as an elementary school teacher, you probably don't want to hear that your former students were confused because they had never used those units of measure within your science lessons.

The reality is that the metric system is not especially difficult to learn if students have multiple opportunities to use it. Teaching how to convert back and forth between measurement systems is not a good use of time and energy. Instead your goal, when it comes to metrics, is that the students learn to switch from one system to the other as easily as some people switch between languages. People fluent in multiple languages don't translate every phrase in their mind—they just speak. Inches and centimeters are simply different units of measure that we'd like students to be able to represent using their hands (can you show an inch with your thumb and finger on one hand and a centimeter with the other hand?). But memorizing the formula to convert Celsius to Fahrenheit really has no place in elementary and middle school science.

Perhaps the biggest barrier to learning the metric system is attitude. In this regard, learning science is similar to learning a second language. If a person feels as if he or she is required to learn, whether it's metrics or Spanish, for example, then the depth of understanding is not as deep. However, if learning the new material is regarded as not only useful but also interesting and valuable, then the rate and quality of learning is greatly enhanced. Students will notice if their teacher is resistant to or afraid of the metric system. Positive dispositions are contagious.

Most of us can remember, as high school or college students, where we lost points on an assignment or quiz because we didn't include proper units of measure. At the time, this rule seemed arbitrary and unfair. After all, you might have thought, if the calculations were correct and the numbers turned out right, then what difference does it make whether the answer includes centimeters or grams or whatever? Perhaps at the root of such disputes is an incomplete understanding of the purpose of the assignment. If our task was to calculate how much hot water would be needed to cool the total volume to 40 degrees, then a number like "120" really has no meaning. Here are parallel examples of insufficient information:

Question: How much baking soda should we add to banana bread?
Answer: 10.

Question: How much fertilizer should be added to the plants?
Answer: 25.

Question: How much of your hair should be trimmed?
Answer: 1.

Does this make the importance of including units of measurement more relevant? This is the same issue that scientists and teachers have with including units. Otherwise the answer is less than incomplete—it doesn't make much sense.

If students are allowed to develop skills at measuring, in multiple contexts and for multiple purposes, then some of the struggles we associate with measuring, including the metric system, will fall aside. Frequent opportunities to conduct measurements, numerous occasions to select among tools for measuring, and abundant instances in which measurements are included within small group and whole class discussions all make measuring a natural part of the students' repertoire of science process skills. As a result, the habit of recording and reporting units of measure will begin in the lower grades and will be significant to the students' efforts. This is much more likely to produce long-lasting understandings than following a rule such as "always include units" and reminding students with red marks and frowning faces on their papers. It is in the light of such experience that the need for units becomes intelligible.

PREDICTING

Uncovering patterns in nature is the goal of the work of scientists. In many ways, finding the patterns makes life easier because many of us take comfort in knowing there is regularity to the universe. Substitute teachers are well aware of students' desire for regularity. They are almost always faced with students who point out "that's not the way our teacher does it." Routines, traditions, and predictability are important to humans.

Predicting involves making a statement that forecasts what will happen in the future. The expectation with predictions is that we will be able to test their accuracy. In this way we create a cycle of pattern seeking: we observe an event, we infer a pattern, we predict what will happen, and then we observe to see if the prediction turns out to be accurate. The goal is not simply to make good predictions but to find patterns that allow us to decide whether our observations, and the inferences we made from these observations, make sense.

Testing Our Understandings Against the World

Predicting is a science process skill we rely on within our daily lives. A wind begins to kick up from the south, and we grab an umbrella as we head out the door. There's a long weekend coming up, so we buy gas in advance because we expect the prices to rise. The professor always has more than enough to keep the class busy until the last minute, so you don't expect there's much chance you'll be dismissed early. We use what we know to anticipate what's going to happen, and we find out whether our prediction is correct.

Beyond the comfort that comes with identifying patterns, there's money and success to be made in making accurate predictions. Stockbrokers and real estate speculators try to predict human actions and reap the financial benefits when they're correct. Police detectives predict criminal behavior, physicians predict the spread of disease, engineers predict the strength of a structure, and teachers predict the behaviors of their students. Predicting is not exclusive to the culture of science.

However, there is a way that predicting is unique to science. When we predict in science, we are not simply making a guess. Whenever we predict in science, we rely on the patterns that we have observed or even measured. Rather than guess what phase the moon will be on your birthday (Why? Maybe you're contemplating a midnight party?), you could use what you know about lunar cycles as a pattern for predicting if the moon will be full that night. When we guess, we don't always have much basis for that statement, but when we predict, we have to have some underlying rationale that is based on observations and previous experience.

FOR REFLECTION AND DISCUSSION

Suppose you were going to introduce the process skill of predicting to students. Before you begin the lesson, you want to share with the students how the process skill of predicting is important. How many professions or occupations can you list in which predicting, and predicting well, is an important aspect of the job? Think broadly, and don't restrict yourself to scientific careers.

Benefits of Making Mistakes

We often use the phrase "we learn from our mistakes" when we've really messed up and someone is maybe trying to make the best of the situation. As a result, most of us don't like to hear the phrase because it usually comes right after we've been told that we were wrong. However, if we distance ourselves from the sting of the phrase, we may recognize its value when it comes to the science process skill of predicting.

Here's a scenario: a teacher is doing a demonstration for the class. The teacher holds a dry paper coffee filter over her head and releases it; the class observes as it falls to the ground. Producing a stack of ten coffee filters, the teacher asks, "What do you think will happen if we compare the falling of one coffee filter to ten—which do you predict will reach the floor first?" Students speculate out loud, and the teacher directs them to write their predictions in their notebooks. And then they observe as the investigation continues.

Suppose a student made a prediction that turned out to be correct, but the reasons used to make the prediction were wrong. How might that individual construe what was observed? His or her expectations were reinforced and, as a result, the understandings of the situation remain incorrect but even stronger than before. In contrast, suppose an individual made a prediction that turned out to be wrong. How might this student respond? We hope she or he would not dismiss what had been observed as some trick. Instead the student would need to reconsider the reasons for the prediction. Ideally the thinking would lead to generating a new pattern, because the old pattern failed to stand up to the test.

Which of these would you prefer to happen in your classroom? Many of us are concerned about the possibility that our students will become confused during our lessons, and we shudder at the notion that they will become discouraged. However, another danger is that students will be highly confident in their knowledge even when they are incorrect. How can we alter the situation so all the students are confident and correct?

COMMUNICATING

The sixth and final basic science process skill is certainly the most broad. Although there are very specific examples of the previous five process skills, along with criteria that sort the proper form from the improper form, communicating is not nearly as restrictive. This doesn't mean that anything qualifies as communicating, but there are many options. The process skill of observing includes more than observing with the eyes—listening, smelling, and touching can also lead to important observations. Similarly, the process skill of communicating extends beyond the use of written text. Students can communicate their observations through their drawings or models, their verbal inferences, and their measurements in the form of a graph. These are examples of the appropriate use of communication as a process skill.

Communicating and the Other Process Skills

It is difficult to imagine how communicating can be separated from the other five process skills. After all, observing would reasonably lead to sharing that information with others. Classifying promotes the recording of properties or the oral explanation of the rules used to divide objects into two groups. Even when the audience for communication is oneself, such as when recording measurements or predictions for later reference, the act of communicating seems tightly bound to the other process skills.

Then why do we list communicating as a separate process skill? One reason is that communicating reinforces the social dimension of the science culture. Scientists rarely work in isolation—they communicate with others. Scientists must gather evidence to combat skepticism. Scientists rely on their writings, sketches, graphs, and so on to help them recognize the patterns in nature. **Communication** as a science process skill is necessarily connected to the other process skills.

In our professional opinion, we should not consider every act of communication by students as an example of their use of this science process skill. Our preference is that teachers connect communicating with another specific process skill before they claim that their students are doing science. In other words, if students are writing or drawing but they are not basing those acts on other process skills, then we argue that their action shouldn't be considered science. Science has its limits, and even though many of us perceive the potential for science all around us, that doesn't mean that everything *is* science—including communicating.

What Qualifies as a Science Lesson

If science has its limits, does that mean that we are opposed to integrating science with other subjects? Absolutely not. However, we foresee problems when a teacher believes the students are doing science when they actually are not. We will devote more space later in this book to curriculum integration. But for right now, we want you to consider this claim:

> Students must be using the science process skills for us to claim that they are doing science.

This provides a clear way to distinguish science from lessons in other subject areas. Too many people associate school science with reading textbooks and doing worksheets—and that's all. We wish to emphasize that science requires students to engage in science through their use of the process skills. We resist any claims that students can learn observing, inferring, and predicting just by using a book. We urge teachers to apply process skills, as much as is feasible, to objects being studied. In most instances, using the process skills with the actual materials is preferable to just reading about those objects. Even though not every science topic can be studied through direct observation (the solar system and dinosaurs come to mind), students should study actual objects and not just pictures and stories.

When students are learning all of their science without using the process skills (such as by reading a chapter and answering questions at the end), it's not clear what label to give to that subject. In such a case, students are becoming familiar with some of the objects of science, or its explanations, but that learning is incomplete. Furthermore, if students are doing a hands-on activity but they're not using science process skills, such as when they follow the steps in a procedure without any sense of purpose, then their work wouldn't qualify as science either. In this case they are participating in actions but not clearly the actions of science. One of the central goals of science is to come to an understanding of the natural world. If students are "doing" but without having a clear purpose, then these actions are not legitimately actions of science.

Let's find our way back to the process skill of communicating. There are many ways in which students can use communicating within the context of science, but it's a science lesson only when the other process skills are involved. Does this mean that students can create posters, present a skit, or perform a dance as part of science? Can they communicate in these ways and still be considered to be doing science? Not automatically—there has to be a clear connection to the other process skills. If the basis of their understandings comes from phenomena they are observing, inferring, classifying, and so on, then what they are doing would legitimately count as the science process skill of communicating. Students' direct experiences with food chains, moon phases, or rock erosion that are translated into creative representations of their understandings should be included in the actions of science.

Communicating and Cultural and Linguistic Diversity

The action of communicating within science is a central one to focus on when teaching science to students from diverse backgrounds and particularly students who are learning English. Closely observing students' efforts to communicate can allow a teacher to discern what a student is learning—if one is vigilant. Communicating is a very important tool for learning, but it requires teachers to move beyond an overdependence on writing as a primary means of communicating. This expanded notion and appreciation of communicating has positive effects beyond building students' vocabulary and fluency.

Having students draw, model, and even role-play ideas allows students who are acquiring basic literacy skills to engage with the content in a meaningful manner and express what they know.

FOR REFLECTION AND DISCUSSION

Communicating can sometimes be regarded as sharing what you know, whether by writing, speaking, or using some other form. But it also has been reported that the act of explaining ideas helps a person learn. How has the process of expressing your ideas, such as when you're asked to "write a reflection," assisted you with clarifying what you know?

We acknowledge that we are setting high expectations for teachers. They must design experiences so that students use their process skills in the quest for scientific understanding even as they learn to communicate those understandings to others. For English language learners and students with cognitive disabilities, the teacher should not stop there. As students communicate ideas through alternative means, the teacher should step in and model for students what their explanation might look like when expressed in a traditional, written manner. This could involve something as simple as the teacher writing a brief explanation on an overhead projector or on colorful charts posted around the room. Through these activities, English language learners are offered another opportunity to acquire English as they are learning science. Also this approach assists students with cognitive disabilities who will be more likely to encode or remember scientific ideas. Teachers need to model more formal means of communicating ideas for students but only after students have arrived at these notions themselves […].

POINT-COUNTERPOINT

There is considerable appeal to the ideas of multiple intelligences and learning styles, although there is a growing controversy. Proponents regard the concept of learning styles as a very useful tool for thinking more broadly about students' capabilities. But others worry that learning styles may actually interfere with high-quality teaching. So in response to the question "How useful is the concept of learning styles?" we provide two contrasting views.

The Problems and Misuses of Learning Styles Information

Marcy Driscoll
Professor of Psychology and Dean of College of Education, Florida State University

"He's a visual learner." "She's an auditory learner." "He's a right-brained dominant learner." When I overhear teachers making statements such as these about the students in their classes, I cringe. Don't get me wrong. These teachers' hearts are in exactly the right place. They know that today's classrooms are very diverse places. Children come to school with a wide range of knowledge and skills, representing different races, ethnicities, and cultures and having had a multitude of different home experiences. There is no question that students have different learning needs, and it is up to the teacher to determine what those needs are and then attempt to meet them.

However, there being different student needs does not mean teachers should use learning styles to guide their instruction. The notion of "learning styles" is so intuitively appealing that it has been transformed into a big business. Websites, reference books, and other sources offer questionnaires to identify learning styles and instructional strategies to accommodate to learning styles. But as much as I support the value in teaching students as individual learners, I have some problems with the learning style movement.

Learning style is defined as the way a learner prefers to take in and process information. Proponents argue that mismatches between student learning styles and instructional strategies can lead to student boredom, problems with motivation, and lack of achievement. Many researchers have experimented to see whether matching instructional style with learning styles improves learning. Most of these efforts have failed to show any advantage of matched instruction over mismatched instruction. Time and again, researchers, using one kind of student characteristic or learning style after another, consistently find no difference between groups. There simply is very little empirical evidence to support the construct of learning styles.

The authors of a recent research article describing a learning styles inventory cautioned about its use, and their comments echo my concerns. The use of learning style inventories doesn't point to particular strengths or weaknesses regarding a student's ability to learn. Too often, learning style information leads to labeling students as certain types of learners. In addition, teachers may use learning styles as a justification for providing students with activities simply because of the supposed preferences. This is not the intended use of learning style inventories, but it seems to happen quite often. Although a learning style framework is meant to encourage teachers to use a variety of strategies, too often it seems teachers restrict their efforts because they perceive some students can learn only through methods that match a particular learning style.

Designing effective instruction should be an important goal for any teacher. I agree. But trying to design effective instruction based on learning style preferences has been proved over and over not to work. Rather, it is more important for all students to be exposed to a variety of ways for representing and processing information. In other words, learning styles should promote a wider range of teaching strategies. The more connections students can create in their memory and the more senses they use during learning, the more durable will be the knowledge and skills they acquire. We do know that research shows that effective instruction consists of three strategies: communicating to the students what they are expected to learn, helping relate the new material to what the students already know, and designing instruction that touches on issues the students care about and see as relevant. These are the guidelines that teachers can follow to assist their students' learning. Knowledge of these principles will help teachers to meet their students' needs. But knowing about students' learning styles will not.

The Usefulness of Learning Styles in Classrooms

Kathy Manning

Middle School Science Teacher in Shaker Heights, Ohio

Do not train students to learning by force and harshness; but direct them to it by what amuses their minds so that you may be better able to discover with accuracy the peculiar bent of the genius of each.

—Plato

As a middle school science teacher, I love the first day of the school year. I welcome each student at the door, smiling and doing the best interpretation of Mary Poppins I can muster. Each hormonally propelled adolescent begins with a clean slate, and I hold no preconceptions about their strengths and weakness. On the first day, every one of my new students is a creative and intellectual genius. But on the other hand, I really don't like the first few weeks of school because I know nothing about my students. Even though everything seems possible on the first day, it takes me awhile to develop a good sense about how much to push and how much to back away. Developing healthy relationships with my students allows me unbelievable freedom. I can convince them that studying is worthwhile, that thinking is something they can do, and that it is good to learn science. But first I must learn who my students really are.

One tool I use for gaining insights about my students is a learning styles questionnaire I have created. I spend a few minutes preparing them for the "test." They groan and roll their eyes, but their attitudes change when I explain how I will use the information the questionnaire will provide. I tell them that the information will be used to help plan activities, assignments, and assessments that will match their areas of strength. Their honest responses to the questionnaire will help make me a better teacher. After they complete the inventory and we score it, we discuss the significance of the scores. For many, I suspect they want to hear the answer to important questions such as "am I normal?" or "am I stupid?" The scores show that each person is different and that this uniqueness is worthwhile and something to build on.

Using my learning styles questionnaire lets my students know that I am interested in helping them to be successful in science. This information also reminds me that my teaching must rely on an array of strategies so each student can succeed. I use a wide variety of teaching and learning strategies I have gathered over the years. I don't fall into the trap of teaching each student according to his or her learning style. Instead, I use a wide range of approaches because I know that one student's success can create a ripple effect. The learning of one individual extends my reach by helping to advance the learning of others.

One interesting discovery I made about using my questionnaire was the reaction by the parents, which is very positive. I believe part of this is because they appreciate that the teacher is willing to try to get to know their child. Rather than acting as if their child was just one of the hundred of students I teach, I reveal a genuine interest in the particular qualities of their child. I use the information when I have parent conferences, and they seem to respond well. I'm not sure that the learning styles information is all that important to them, although they are intrigued to fill in the questionnaire themselves. What I think is more powerful is the idea that a science teacher is not only interested in the content. They seem relieved and comforted by my interest in my students.

Finally, by far one of the strongest reasons I've discovered for using the learning styles strategies is that they help identify many reading and learning disabilities or challenges that would otherwise go undetected by me as an untrained learning disability specialist. As a science educator, I am confronted daily with the need to cover more and more content with greater depth. Knowing my students and their ways of learning gives me the insight I can use to provide them with the best possible approach. Though I've never "seen" a million dollars, I know it exists (just not in my bank account). Though the research may not yet prove that learning styles is a perfect education theory, I have other pieces of evidence that support its use. I see with my own eyes the effect it has on my students, the attitudes of the parents, and my appreciation for their uniqueness. I think Plato would approve.

PROCESS SKILLS WITH ENGLISH LANGUAGE LEARNERS

As part of our commitment to teaching science to all students, we must pay attention to the particular learning needs of those students whose first language is not English. At the moment we are emphasizing the teaching and learning of the science process skills. Perhaps a natural question would be whether there is anything especially important to keep in mind when providing process skill instruction to English language learners. It might be tempting to suggest that there is nothing different we should do when using process skills with these populations. However, we are reluctant to propose that this is a one size fits all situation. The reason for this caution is that it may lead us to disregard the need to differentiate science instruction at all. To avoid that fate, we will turn to research studies concentrating on subject matter learning in multilingual classroom settings.

The Center for Research on Education, Diversity, and Excellence (CREDE) originated with the goal of guiding those students who face the challenges of culture and language differences, race, and poverty to attain their maximum academic potential. It is fortunate that the CREDE included in its work the goal to provide very useful information about science education to researchers and teachers (e.g., Warren & Rosebery, 2002). The CREDE organization developed a set of *Five Principles for Effective Pedagogy* toward the goal of educational excellence for diverse students. We will use these five principles as a reference point in our discussions of the science learning of diverse student populations. We briefly introduce all five principles here, and for our immediate purpose we will focus on the first principle only. The other standards will be examined at appropriate places in other sections of this book.

Principles for Effective Pedagogy

The CREDE group has organized a quintet of instructional approaches for teaching diverse student populations how to achieve educational excellence. More than simply techniques, the five principles describe a set of guiding approaches. Although some people might claim these standards are indicators of "just good teaching," our perception is somewhat different from this ordinary assertion. As we illustrated early in this text, there are inequities in our current educational system that, in pure and simple language, are unfair to some students. Although the exact causes are not definitive and obvious, there exist achievement gaps in science corresponding to gender, cultural background, native language, and family economic factors. In light of these disparities, it seems a certain level of urgency and earnestness needs to be assigned to the identification and implementation of strategies that will reduce, if not eliminate, such inequities. For those who prefer to treat the five principles as simply recipes for good teaching, we still support their efforts to employ the principles in any classroom. Yet for those students who, both historically and currently, have been inadequately served by mainstream education, the five principles represent an educational imperative. Either way, these *Five Principles for*

Effective Pedagogy (Dalton, 1998) are worth learning by those who teach science in the elementary and middle school grades.

1. Joint Productive Activity: Teacher and Students Producing Together
2. Developing Language and Literacy Across the Curriculum
3. Making Meaning: Connecting School to Students' Lives
4. Teaching Complex Thinking
5. Teaching through Conversation

Each principle constitutes a suite of instructional approaches and embedded assumptions, and it would be too much to try to describe all of them here. Instead, we will return to each principle in separate chapters to interweave the complex ideas inherent in each principle with the relevant aspects of science teaching and learning.

Joint Production of Understanding

The first principle of effective pedagogy for diverse learners describes a situation in which the teacher purposefully works side by side with the students to understand scientific ideas (see Table 13.1). In this case, the teacher is seen as the expert learner and the student is an apprentice to those ways of thinking, doing, and knowing (Rogoff, 1990). By teacher and students working to resolve a practical problem or achieve a shared goal, the work becomes a joint effort. The beauty of this principle is the creation of a common experience within the context of the science classroom wherein the teacher and students are co-constructing scientific understandings. Despite differences in individual cultural backgrounds, the joint work not only assists students' learning of science but also can support communicating skills and encourage the appropriate use of academic language (LAB at Brown University, 2002). A distinguishing feature of this principle is recognizing the teacher as an active member of and participant with the community of science learners.

The implications for this principle become somewhat clearer when we examine actions teachers could take that would support the joint production of understanding. On the surface, this principle may seem to simply reinforce the idea of group work. However, there is more nuance at play as the teacher joins in—not just in a playful sense but in a genuine engagement with the problems, questions, or activities. Furthermore, by engaging in productive activities, the teacher creates a mechanism that supports language development, all the while in the context of authentic activities. In other words, in the midst of doing science, the teacher is modeling more formal ways of communicating that students can perceive as an example they can follow. As a result, students' language use becomes richer and more sophisticated.

TABLE 13.1: Joint Productive Activity and the Appropriate Teacher Actions

The teacher plans learning activities that promote collaboration toward a shared goal.

The time allotted for a joint productive activity is sufficient to complete the tasks.

The seating of people encourages individuals to work and talk together.

The teacher joins in the activity with his or her students.

Grouping occurs in a variety of arrangements: friendships, mixed academic abilities, blended language fluencies, and matched individual personalities—in short an array of efforts to support interactions.

The teacher instructs the students about the social skills required as part of working in groups.

The teacher structures the access to materials and monitors time (e.g., efficient transitions, allowance for cleaning up, etc.) to facilitate joint productive activity.

Note: Adapted from Dalton (1998).

Let's bring this idea of joint productive activity back to our preceding discussion of the basic science process skills. As further support of our efforts to discard the myth of "teachers as experts," engaging with students in science activities seems appropriate. Actually, for a teacher to take a seat with a group of students as they observe, predict, measure, and so on with a group of interesting objects (pinecones, seashells, cereal box labels) not only is OK but represents good teaching. To the uninitiated, the thought of a teacher joining in with his or her students during a hands-on activity seems charming. But the savvy educator will recognize this action as much more substantive. In talking with students about their work, the teacher models effective communication, encourages the voicing of ideas, and creates a forum in which all children, and especially English language learners, can enhance their facility at oral communication.

By now you should be developing a sense about the power and utility of infusing science teaching with the basic science process skills. A key theme is the necessity of giving students many opportunities to apply these process skills firsthand through their study of actual objects. But here we are not talking about activities that the teacher steps back and allows the students to perform; rather the first principle of effective pedagogy in terms of diverse student populations is the value of the teacher joining in during productive science activities. There is even clear evidence not only that a focus on science process skills is appropriate for students with various cognitive disabilities but also that such an emphasis is actually beneficial to the learning of such students (Mastropieri & Scruggs, 1992). Before concluding our formal consideration of the basic science process skills, we will devote some attention to the way the process skills are inherent within the *National Science Education Standards*. Also, before the chapter is over, we will address the concerns sometimes expressed about teaching elementary and middle school science using the science process skills.

PROCESS SKILLS AND SCIENCE STANDARDS

With all the emphasis placed on the science process skills, you might expect that they would have a prominent place in standards documents. However, if you obtained a copy of the *National Science Education Standards* (National Research Council, 1996), you might be surprised to not find "process skills" in the index. Process skills first became a substantial feature of elementary and middle school science and teaching learning forty years ago. The reason for the shift four decades ago was the response to the Russian launch of the Sputnik satellite. This event, which signaled a threat to America's scientific, educational, and military superiority, was used to leverage changes in science teaching, especially through the development of an innovative science curriculum.

One common goal of the elementary and middle school science programs of the 1960s was to shift from traditional, textbook-based science materials to more active, hands-on activities. The reformers imagined that science instruction should not occur with students sitting quietly at their desks and taking turns reading from their textbooks. Instead, the students would be working with equipment, talking with each other, and experiencing science in ways consistent with how science was actually done. This shift in student activity was accompanied by a shift in the science curriculum. Books are an ideal way to push scientific vocabulary. However, activities are not as obviously linked to vocabulary enrichment. Within this new approach to science instruction, an emphasis was placed on the science process skills.

One of the more well-known curriculum projects of the 1960s and 1970s was "Science: A Process Approach," also known as SAPA. SAPA was developed under the aegis of the American Association for the Advancement of Science (AAAS) and ultimately contained more than one hundred activities within its K–6 program. The structure of the activities, the presence of a considerable amount of equipment, and the absence of text materials sent a clear message to teachers: learning scientific facts in elementary school was not nearly as important as comprehending the processes used by scientists. This was tightly connected to the belief that science learning should happen by actually doing science, not simply reading about it. For a variety of reasons, partly financial but also political, the SAPA program was unable to sustain itself. In the 1980s, several researchers (Bredderman, 1983, 1985; Shymansky, Kyle, & Alport, 1983) evaluated previously published studies of SAPA and other programs of the post-Sputnik era and found that these materials were at least as good at helping children learn as teaching science from textbooks alone, and in most instances they were better. Measures of the students' creativity and science process knowledge were higher for students who used this activity-based program than for students who did not use it.

Process Skills: In Context and with Content

What is the reason for this history lesson? In part it shows how science teaching and learning is continually changing. This means that the way you approach teaching science at the start of your career will not be the same as the way that you were taught science in elementary and middle school. In particular, standards and accountability are much more influential nowadays than even just a few years ago. One lesson is to recognize that education is in a constant state of flux. The other lesson is to prepare yourself for those who might question the emphasis placed on science process skills. It seems possible that the resistance to building science teaching around process skills is an overreaction to the emphasis given to process skills in previous generations of science curriculum. However, that's only our speculation. Closer to the truth is probably the view that students really can't and shouldn't learn science by mastering the process skills in isolation from science content. In particular, the authors of the *National Science Education Standards* have made it clear that they see science learning as necessarily combining the learning of science process skills with science content:

> The new vision [as presented in the *Standards*] includes the "processes of science" and requires that students combine processes and scientific knowledge as they use scientific reasoning and critical thinking to develop their understanding of science. (National Research Council, 1996, p. 105)

This desire is shown in a diagram (see Table 13.2) within the *National Science Education Standards* in which the old ways of science teaching and learning are indicated as being something to which we should be giving less emphasis (the left-hand column) and the preferred ways of going about science teaching are indicated as things that deserve more emphasis (the right-hand column).

There is much in this diagram that we support. Specific to our immediate concern is the third line down that highlights a shift we endorse: science process skills should be used within the context of learning significant science content. However, we feel the next line down in the table is perhaps too idealistic by claiming that individual process skill use should be de-emphasized. First of all, it makes good educational sense to present students who are new to learning science with clearly defined process skills. We can see a danger in students using only one process skill at a time. But we aren't fearful that you or most other teachers are going to take such a narrow path. Also the ways in which students should be expected to learn science do not necessarily have to be identical with the ways in which scientists do science (Millar & Driver, 1987). Perhaps in previous generations, where science education was seen as a site for nurturing future scientists, this would be appropriate. However, this perspective is much too limiting. Our new challenge is to make scientific literacy a possibility for all students. Viewing the science process skills as an integral feature of the actions of the scientific culture, although not as all there is to science, we have become convinced that teaching with an eye toward science process skills is an appropriate entry point for beginning elementary and middle school teachers.

TABLE 13.2: Changing Emphases for Inquiry as Identified within the *National Science Education Standards* (National Research Council, 1996)

LESS EMPHASIS ON	MORE EMPHASIS ON
Activities that demonstrate and verify science content	Activities that investigate and analyze science questions
Investigations confined to one class period	Investigations over extended periods of time
Process Skills Out of Context	**Process Skills In Context**
Emphasis on individual process skills such as observation or inference	Using multiple process skills—manipulation, cognitive, and procedural
Getting an answer	Using evidence and strategies for developing or revising an explanation
Science as exploration and experiment	Science as argument and explanation
Providing answers to questions about science content	Communicating science explanations
Individuals and groups of students analyzing and synthesizing data without defending a conclusion	Groups of students often analyzing and synthesizing data after defending conclusions
Doing few investigations to leave time to cover large amounts of content	Doing more investigations to develop understanding, ability, values of inquiry, and knowledge of science content
Concluding inquiries with the result of the experiment	Applying the results of experiments to scientific arguments and explanations
Management of materials and equipment	Management of ideas and information
Private communication of student ideas and conclusions to teacher	Public communication of student ideas and work to classmates

We have found that the science process skills serve as a very important way for new teachers to learn about science teaching. Everyone needs to start learning somewhere. We accept the claim that science process skills alone are not enough, and we are not saying they are enough. But every new teacher needs a place to begin, and any book about how to teach science must start somewhere. From our professional experiences, and from the feedback we've received from former students who return to us for graduate studies, the science process skills serve as an excellent jumping-off point. The list of basic science process skills is fairly easy to remember (you can invent your own mnemonic device if you want to memorize all six), and it is not especially difficult to find opportunities to incorporate them into almost any hands-on science activity. Even though you will discover that science should not simply begin and end with the process skills, they are a powerful foundation on which you can build the other knowledge and skills for teaching science.

We suggest that you put the science process skills at the very core of your science teaching repertoire. We will show you how to add another layer around this nucleus: how to plan lessons. On top of this we will add layers to help you improve your use of discussions and skill at asking questions. Later you will develop more knowledge about other curriculum materials, assessment strategies, and educational technology. At the middle will always be the science process skills—they aren't enough on their own, but there is so much we can build around them once you recognize their value, simplicity, and power. Did the authors of the *National Science Education Standards* exaggerate their concern about the presence of the science process skills? Perhaps, but because they did so many other things well in that document, we won't complain over the neglect of the process skills.

THE SCIENTIFIC WORLDVIEW

The culture of science represents a way of thinking, knowing, and being in the world that all students should learn. Teaching science to children is powerful because you can provide your students access to the culture of science—and all the political, monetary, societal, and other types of power it can provide. We have presented the six basic science process skills as components of the culture of science that are accessible to all students from kindergarten to the higher grades. Knowing how to use the science process skills can be an important framework for assisting students to engage in genuine scientific study. When process skills are used in the context of scientific information, they will support the students in their acquisition of scientific content knowledge. In short, the process skills provide all students with access to the scientific worldview.

One approach used by science methods instructors to drive this home is to assign their pre-service teachers to participate in a moon investigation (e.g., Abell, George, & Martini, 2002). When we use this assignment in our methods course, we ask our students to maintain a "moon diary" for one month. We ask them to draw the moon's shape each night, note its location, pose questions that come to mind, and keep track of their thinking along the way. At the end of the four weeks, they are to write a summary paper in which, among other things, they report their use of the science process skills. They discover, often to their surprise, that they have used all six science process skills. For example by characterizing the moon's shape as oval or more like a watermelon rind, they have classified. Through classroom discussions, they also develop conceptual understandings of moon phases, eclipses, and other fairly abstract ideas. As a consequence, these future teachers develop a scientific worldview.

EXPERIENCING THE SCIENTIFIC WORLDVIEW

Recognizing that it is quite limiting to only talk about a scientific worldview, we are providing you with an activity that requires only very basic materials. The pendulum was a tool that Galileo used during his studies of objects in motion. Its simplicity makes it an excellent object for reinforcing a worldview that is consistent with the culture of science. You can perform this activity by yourself, but we encourage you to find someone else to collaborate with on this project.

A pendulum is an interesting object to investigate in collaboration with someone else. To make a pendulum you need a piece of string (at least thirty centimeters long) and some object you can tie to it that is heavy enough to cause the string to hang straight. Tie the object to one end of the string, hold the other end of the string so it is well anchored and won't move, and push the object to make it swing from side to side. Don't contribute any force to the pendulum once it starts swinging. Once released, the weight should have all the impetus needed to keep swinging. If your anchor hand twitches and wiggles, then the pendulum will slow down prematurely. Keep the anchor steady. Now find a way to keep track of time while counting the number of swings the pendulum makes.

Count the number of times your pendulum swings in fifteen seconds. Try it a few times, without changing where you hold the string or anything else, until you obtain consistent results. Because you now have some experiences, you are going to predict (and not guess) what will happen to the pendulum when you change the conditions.

1. If you change where you hold the string so the anchor point is closer to the weight, do you predict that the number of swings in fifteen seconds will increase, decrease, or stay the same? What does your coworker predict? Use the materials to check your predictions.
2. What will happen if you change the height from which you release the weight? In other words, if you raise the weight so the angle the string makes is larger, will the number of swings increase, decrease, or stay the same? Does your coworker's prediction match with yours? After discussing the possibilities and deciding on your prediction, see what happens.
3. Find another weight you can add to the string. What do you think will happen if you keep everything about the pendulum the same (same length, same angle, etc.) but have a heavier weight on the end: will the number of swings increase, decrease, or stay the same? Does your coworker agree with you? Once again, see what you find out by actually testing the pendulum.

What you notice and how you attempt to explain these facts reflect your ability to observe and infer. When you wonder about the behavior of the pendulum and are puzzled by the data, then you're experiencing the habits of mind of curiosity and skepticism.

It isn't uncommon for the investigation of pendulums to produce surprising results. There is something counterintuitive about the results. We often expect more factors to be the cause of differences in the swing rate of a pendulum than we end up discovering. When our expectations are different from the experimental results, we find ourselves in disequilibrium. In a way, the puzzling results are similar to the actions of the pendulum. Essentially, you perturb the pendulum system by lifting the weight and releasing it. In much the same way, having data in front of you that doesn't align with your expectations is also perturbing. The pendulum's response to being perturbed is to swing from side to side until it stops in a stable position. Likewise when your thinking is perturbed, you may find yourself puzzling over an explanation and perhaps developing or acquiring an explanation that puts you back into balance. This is when your mind is going through accommodation. Along the way, you've veered into the culture of science and come into contact with elements of the scientific worldview. Perhaps you've even noticed that this experience has given you renewed interest in what it's like to learn science.

CHAPTER SUMMARY

- Measuring is a special form of observing through the use of a tool that extends our senses. Measurements represent quantitative observations and are based on accepted standard units.

- Proper use of various measuring tools and proficiency with the metric system are largely a matter of familiarity. The more frequently these tools are used, the simpler they seem to be.

- Predicting allows students to test their ideas by forecasting what will happen. By posing predictions and then testing their accuracy, students and scientists can begin to uncover patterns. On those occasions when predictions turn out to be incorrect, the cognitive disequilibrium that is produced becomes a motivation for understanding what was really happening.

- Communicating is directly connected to the other science process skills. Whenever communicating occurs without being tied to any science process skills, the communication is being used for some purpose other than science.

- The process skills not only represent how science is done (what we have taken to calling the "actions of science") but also become tools for enhancing student learning. Measuring helps to focus and refine observing, predicting encourages the seeking of patterns, and communicating leads to the clarification of individual's understanding. In each case, the ideal situation is one in which these process skills are being used with actual objects and not simply representations (pictures or texts) of objects.

- The science learning of all students, but especially those who are not yet fluent in English, benefits by the teacher becoming part of the learning community. The shared participation of the teacher and students working jointly as everyone applies the process skills strengthens understandings of content and language.

- Science is one approach for understanding our world and should not be regarded as superior to any other worldview. Nevertheless, by becoming fluent within the culture of science, students have access to forms of power that can give them more control over their lives.

KEY TERMS

Communication: the broadest of the scientific process skills, involving the transmission or exchange of ideas through a wide range of media, including verbal means, written text, drawings, and graphs.

Measuring: a science process skill in which a person makes a quantitative determination of the physical world through the use of a standard comparison tool such as a ruler.

Predicting: a scientific process skill, involving making a statement based on some scientific knowledge claim that forecasts what will happen in the future. Predictions are often thought to be a useful means for testing the utility of scientific knowledge claims.

Qualitative: a qualitative observation is a form of observation in which numbers are not employed.

Quantitative: a quantitative observation is a form of observation that incorporates the use of numbers such as by counting or measuring.

SUGGESTED READINGS

Hoffman, J., & Strong, J. (2002). Electric connections. *Science and Children, 40*(3), 22–25.

These two teachers describe the use of predicting and the many other aspects of basic electricity. What is key to this collection of activities is that the students are prompted to evaluate the accuracy of their predictions once the materials are put into use.

Phillips, S. K., Duffrin, M. W., & Geist, E. A. (2004). Be a food scientist. *Science and Children, 41*(4), 24–29.

Teachers reinforced the connections between math and science with an emphasis on food preparation with their fourth and fifth graders. The extended unit of food science proved to be an excellent way to build students' skills at a wide array of measuring approaches.

REFERENCES

Abell, S., George, M., & Martini, M. (2002). The moon investigation: Instructional strategies for elementary science methods. *Journal of Science Teacher Education, 13*, 85–100.

Bredderman, T. (1983). Effects of activity-based elementary science on student outcomes: A quantitative synthesis. *Review of Educational Research, 53*, 499–518.

Bredderman, T. (1985). Laboratory programs for elementary science: A meta-analysis of effects on learning. *Science Education, 69,* 577–591.

Dalton, S. S. (1998). *Pedagogy matters: Standards for effective teaching practice.* Washington, DC: Center for Applied Linguistics.

LAB at Brown University. (2002). *The diversity kit: An introductory resource for social change in education.* Providence, RI: Brown University.

Mastropieri, M. A., & Scruggs, T. E. (1992). Science for students with disabilities. *Review of Educational Research, 62,* 377–411.

Millar, R., & Driver, R. (1987). Beyond processes. *Studies in Science Education, 14,* 33–62.

National Research Council. (1996). *National Science Education Standards.* Washington, DC: National Academy Press.

Rogoff, B. (1990). *Apprenticeship in thinking: Cognitive development in social context.* Oxford, England: Oxford University Press.

Shymansky, J. A., Kyle, W. M., & Alport, J. M. (1983). The effects of new science curricula on student performance. *Journal of Research in Science Teaching, 20,* 387–404.

Warren, B., & Rosebery, A. S. (2002). *Teaching science to at-risk students: Teacher research communities as a context for professional development and school reform.* Santa Cruz, CA: Center for Research on Education, Diversity, and Excellence.

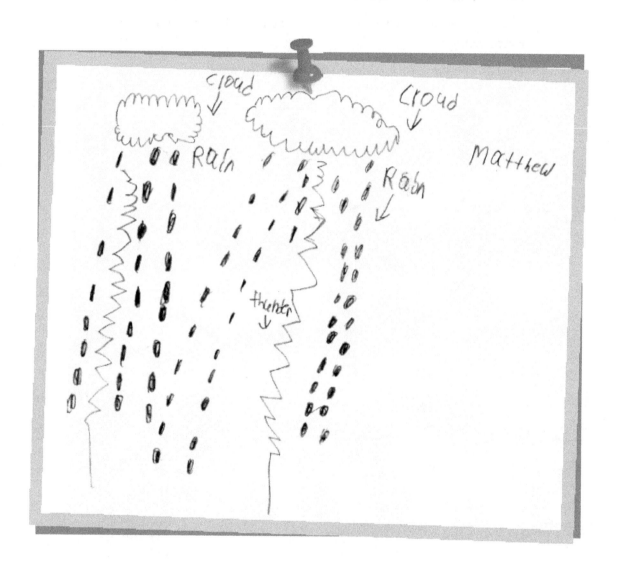

SECTION IV — QUESTIONS TO PONDER ...

1. How do I use brains-on strategies to engage learners and foster learning?
2. How do I foster reasoning and deliver science instruction for ELL students?
3. Why do I need to consider social justice when thinking about science education in a rural setting?
4. How do I best assess my students and measure learning?